The Geography of Names

The human urge to name is part of a fundamental process of sorting the world into manageable categories that stretches from prehistoric to modern times. This is extended in the contemporary era with the advent of internet-based tagging, classification, and mapping systems, including new neogeographical tools restructuring what it means to name in the modern world.

This book explores these developments to provide a cutting-edge theoretical and critical analysis of current debates within toponomy. It examines naming practices as primary tools for making sense of complex life-worlds and livelihood systems, arguing that the concept of the name is inherently spatial and visual in nature. It positions itself at the intersection of traditional approaches to toponymy, which focus on cartographic and technical aspects of place-naming, and critical approaches, which examine political aspects of place-naming practices. It opens up new spaces of enquiry beyond these two perspectives to explore the idea of geographical naming as applied to the world at large, focusing on names as referents applied to a range of phenomena including not only places but also people, objects, and abstract ideas. This book provides a naming toolbox that is at once philosophical, religious, political, practical, and scholarly, with examples from Canada (including indigenous communities), the Middle East, and the United Kingdom. It will be of great interest to those working in geography, cartography, anthropology, and cultural studies.

Gwilym Lucas Eades is Lecturer in Human and Environmental Geography in the Department of Geography at Royal Holloway University of London, where he is also on the Information and Communication Technologies for Development (ICT4D) management board and director of the GeoVisual Methods Lab (GVML).

Routledge Studies in Human Geography

This series provides a forum for innovative, vibrant, and critical debate within Human Geography. Titles will reflect the wealth of research which is taking place in this diverse and ever-expanding field. Contributions will be drawn from the main sub-disciplines and from innovative areas of work which have no particular sub-disciplinary allegiances.

For a full list of titles in this series, please visit www.routledge.com/series/SE0514

The Geography of Names

Indigenous to post-foundational

Gwilym Lucas Eades

Routledge
Taylor & Francis Group

LONDON AND NEW YORK

First published 2017 by Routledge

2 Park Square, Milton Park, Abingdon, Oxfordshire OX14 4RN
52 Vanderbilt Avenue, New York, NY 10017

Routledge is an imprint of the Taylor & Francis Group, an informa business

First issued in paperback 2020

British Library Cataloguing in Publication Data
A catalogue record for this book is available from the British Library

Library of Congress Cataloging in Publication Data
Names: Eades, Gwilym Lucas, 1969- author.
Title: The geography of names : indigenous to post-foundational /
Gwilym Lucas Eades.
Description: New York, NY : Routledge, 2016. | Series: Routledge studies in human geography | Includes bibliographical references and index.
Identifiers: LCCN 2016011426 | ISBN 9781138885172 (hardback) |
ISBN 9781315715636 (ebook)
Subjects: LCSH: Toponymy. | Names, Geographical.
Classification: LCC G100.5 .E34 2016 | DDC 910/.014--dc23
LC record available at https://lccn.loc.gov/2016011426

ISBN: 978-1-138-88517-2 (hbk)
ISBN: 978-0-367-66820-4 (pbk)

Typeset in Times New Roman
by Taylor & Francis Books

Contents

Illustrations

Figures

Table

Acknowledgments

I would like to thank Klaus Dodds who, as my mentor and director of the Politics, Development and Sustainability research group at Royal Holloway University of London, provided support, both moral and practical, throughout the writing of this book. Alasdair Pinkerton, Katie Willis, and Varyl Thorndycraft, also at Royal Holloway, offered support with comments on proposals, writing, and landscape, respectively. Dorothea Kleine has been a steadfast supporter of this project through its promotion in affiliation with the Information and Communication Technologies For Development (ICT4D) group at Royal Holloway.

Faye Leerink (Taylor & Francis), an early champion of this book, has watched the project evolve from its formative stages in an informal chat at my office through to the completed manuscript and beyond. I must thank her for her steadfast belief in the worth of a book on geographical names, and its place among other titles examining place names.

Simon Dalby (Wilfred Laurier) and Philip Conway (Aberystwyth) participated in some of the evolution of *The Geography of Names* during a session at the Royal Geographical Society (RGS) meeting at Exeter in 2015. That session included early draft material of sections on Thomas Hardy's Wessex and on the Anthropocene. Our discussion at the RGS on determinism and possibility has influenced the final shape of this book.

Thomas Thornton (Oxford) and Brian Thom (University of Victoria) both helped think through possibilities for theorizing new kinds of toponymy in various discussions about research possibilities on the west coast of Canada and the United States. Meetings both in Oxford and on Vancouver Island were valuable opportunities to discuss formative and developing ideas for this book.

Dr Kate Distin urged me to explore rogation practices, including beating the bounds. A meeting in Oxford was highly productive and inspired a good deal of thinking in new directions without which this would be a much duller work. Dr Distin's books *The Selfish Meme* and *Cultural Evolution* have served as models for my own work since the very early days of my doctoral dissertation at McGill University (where, on Milton Street, I first, and most serendipitously, came across her work in a used bookstore).

I would also very much like to thank Ludger Muller-Wille for inspiring me to travel to northern Quebec to explore Inuit toponymies. His work led me to contact the Avataq Cultural Institute in Montreal, eventually leading to consulting work with Strata360 and Avataq and extensive travel to complete the Nunavik Inuit Land Claims Agreement offshore alongside toponymic verification work.

Some of the work in this book would not have been possible without the warm and generous support of the people of Wemindji Quebec, a Cree community in eastern James Bay, Canada. The commemorative route and performance of *kaachewaapechuu* each year in remembrance of the ways and stories of the elders still serves as a primary inspiration for the idea of memetic transmission of place-based information and, by extension, the idea of a geographical name-tracking-network that forms the core theoretical contribution of this book.

Amy O'Donohoe at the London Library was most helpful in procuring a comprehensive literature review of the library's holdings on toponymy and philology (the latter tucked into the wonderful mezzanine level). The Bodleian Library was the source of materials on urban and rural rogation and beating the bounds rituals and practices. The British Library holds a large-scale map of Egham, Surrey, which formed the basis of some early explorations into modern-day rambling rituals. The Senate House Library in central London provided some valuable resources, including a source on Mayan place names. Royal Holloway's Bedford Library procured several interlibrary loans in support of this book. The School of Oriental and African Studies in central London also held several sources holding valuable insights for the present work.

Friends, family, colleagues, librarians, and elders whose knowledge is embodied in their stories, movements, and ancestral links to past places and their names – all were essential support networks in the writing of this book, and I cannot thank you enough for your inspiration and material support.

1 Geographical naming and necessity

Why names?

On 15 April 1912, the Titanic sank, taking with it the boat's 'star corpse,' Charles Melville Hays, and changing the fate of Prince Rupert, British Columbia (BC), a small northern port city near the Alaskan panhandle, for ever. 'C.M.H.' (as his monogrammed gold watch informed those who found him) was "an American-born Montreal resident who, as president of the Grand Trunk Pacific Railway, was perhaps the most famous railway executive on the continent" (Hawthorn, 2012). For Charles, as Wikipedia informs us, had big plans to see 'Rupert' (as locals call Prince Rupert) become BC's leading deepwater port. However, due to the unforeseen disaster of the sinking of the Titanic taking out 'C.M.H.,' the leading visionary, it was Vancouver that would go on to become BC's premier location and most populous city, as well as the economic driver for the province. Contingencies such as these, and belief in the power of places and their names to produce possible worlds, are the subject of this book.

This book explores the following question: what is the nature of the geographical name and how (and where) has it changed over time? To answer this question, I take a Kripkean-evolutionary and Wittgensteinian approach.

The Geography of Names is Kripkean in the following three senses. First, it explores possible worlds of named places. For example, Vancouver is the largest city, and principal port, in British Columbia, the westernmost province in Canada. What if, as some posited as possible for a time, Prince Rupert had, some several hundred kilometers north of Vancouver, become the largest and/ or principal port city in British Columbia? Would it have had any effect on the geography of British Columbian place names? Only by positing geographical names as relationally linked and historically contingent artifacts built through cultural place-naming practices does this question about Vancouver and Prince Rupert, BC, begin to make sense. Hanna and Harrison's idea of the name-tracking network (Hanna and Harrison, 2004), or NTN, is useful here because powerful individuals took action that affected the economic power of the cities of Prince Rupert and Vancouver, in turn affecting their relative sizes and the regional 'weight' with which each is associated in the province of

British Columbia. The northern part of the province is less of a 'powerhouse,' with seasonal labor and an economy dependent upon tourism; while the southern part, and the lower mainland where Vancouver is situated, is the economic driver of the province as a whole. This outcome represents just one (realized) outcome among many other potential outcomes, or possible worlds in a Kripkean sense, in which Prince Rupert could have become the economic 'number one' for the province. Consider that it has a deepwater port and also much better access to Asian markets than the southern port, and the idea of historical contingencies becomes more salient. Prince Rupert is in many ways a more suitable site for a 'capital' of BC, but due to a contingent event (the Titanic sinking), one possible world was elided. My hometown of Terrace is located just down the road. This piece of information has relevance for the focus of the opening paragraphs of this book and the subsequent chapter sections that focus on Canadian and British Columbian indigenous groups.

Second, this book examines retroactive senses of the 'rightness' (propriety) of place names. In Kripke's (1981) language, there is an *a posteriori* necessity to the meanings and senses we collectively and individually give to names. It is the collective and societal senses of place-naming practices that give the resulting names such a sense of necessity. The names Paris and London make sense to English speakers and color their sense of both those places in terms of aesthetic qualities and other attributes associated with them. But what precisely does 'Londres' mean to the French speaker? We explore this issue further below; suffice it to say here that it is not simply a matter of translation (though it is also that), but a matter of what the speaker comes to associate with the name in each language. Kripke (2011) has pointed out in his paper "A Puzzle About Belief" that contradictory beliefs about geographical names are possible. To best sort out and resolve contradictions arising from beliefs about places (e.g. their beauty or lack thereof), Kripke suggests that we focus on names and indeed on language. We will have much more to say about belief below, especially in chapter 3, which covers religion and geographical naming; but belief comes into play in chapters 2 and 4 as well, as indigenous naming systems (the subject of these two chapters) cannot be extricated from belief systems. Chapters 2 to 4 are thus ethnogeographical in the sense that they examine beliefs about geographical names, or those associated with them, from the perspectives of the cultures doing the naming (Eades, 2015a).

Third, a Kripkean framework posits both baptism (an original naming moment) and communication chains as necessary to any evolutionary theory of place-naming. This assertion is forward-looking, in a theoretical sense, because it anticipates an evolutionary outlook on place names. However, this outlook was implicit in Kripke's own work and in that of his successors, including, importantly, Berger (2002) whose insight into the workings of communication chains and name transmission are discussed at more length in the conclusion, chapter 7. Suffice it to say here that it is somewhat obvious that certain places, named within living memory—perhaps existing at a frontier or leading edge of development—have newer names than others. Other places' names are so old that the original impetus or logic of the naming has been lost. We posit that the older names did have an original (or baptismal) moment of name bestowal, and that these names are still valid tens, hundreds, or even thousands of years later despite radical change in the originating language, name pronunciation, or even the meaning of the words of the name. What matters in such cases is that the chain (of communication) remains unbroken, for here an evolutionary process can be said to be at play (Richerson and Boyd, 2005). In other words, a thing cannot be said to evolve if its link to the past has been severed. This applies to anything that can be said to evolve, whether a species, an individual, a cultural practice, a belief, or, indeed, a geographical name. It is in the sense of being an evolutionary 'object' that the geographical name can also be theorized as 'memetic' in nature. Memes and memetics are covered in more depth in chapter 4.

The Geography of Names is evolutionary in the following three senses. First, communication chains, along which place-name information is transmitted (Eades, 2015a), change, or evolve, through time. Linguistic variations introduced into the chains (due to drift or colonization) sometimes 'stick' (become permanent)—producing a new or dominant variation of the name. We know from Monmonier's (2006) work that colonial place-naming practices in particular can have an almost viral character, with names turned into memes (durable cultural constructs that travel across both time and space). Monmonier has demonstrated that joke, racist, eponymous, and strange names abound in North America where colonizers, settlers, explorers, and governments took it upon themselves to name things after themselves, their travels, and their whimsies.

Second, place names are theorized as cultural-material artefacts and, as such, are both ethnographic and archaeological, verifiable empirically. It is a verifiable fact that Easthampton refers to a real place that no longer exists. It is similarly verifiable that Atlantis is a fictional place for which belief (both past and present) in its existence is a fact (Plato, 2008). It is the aspect of belief that Kripke made into a puzzle, and it serves in the construction of ethnogeographic facts about places, i.e. those in which belief plays a significant part. Beliefs about places and their names are verifiable facts, as are the names (and their etymologies) themselves.

Third, this book uses cultural evolutionary theory (Boyd and Richerson, 1985; Richerson and Boyd, 2005; Cavalli-Sforza and Feldman, 1981) to construct the premises that, just as places themselves (as cultural material objects of belief and fact) evolve, so do the communication chains caught up in the transmission of their names. Furthermore, it uses a 'unit'-based theory (i.e. one that posits memes as those units) of cultural evolution first made explicit by Dawkins (1976 and 1982), championed and given more nuance later by others (Aunger, 2002; Distin, 2005; Blackmore, 2000), and contested again by many others (Lewens, 2015; Jablonka and Lamb, 2005). The justification for this theoretical move (i.e. use of cultural evolutionary theory) is, furthermore, premised upon the very unit-like nature of place names. But I also posit (though do not prove) that cultural and social life as a whole contains particulate and overlapping sets of beliefs, practices (Runciman, 2009), and indeed names (defined in at least two senses, as we explore below in more depth).

The Geography of Names is Wittgensteinian for the following reasons. First, it uses theories inspired by Wittgenstein. Most important in this regard is Hanna and Harrison's (2004) *Word and World*, which is like a theoretical touchstone for the present work. Specifically, its idea of a name-tracking network is modified here into a geographical name-tracking network (or geo-NTN) for tracking traces, utterances, inscriptions, and devices serving to propagate and evolve (adapt) place-name communication chains. Hanna and Harrison's work is explicitly influenced by the late Wittgenstein (2009) of *Philosophical Investigations* in the sense that it rejects positivist determinations of place-naming in favor of practices associated with belief. I thus consciously avoid philological methods such as tracking linguistic changes through the use of various word particles in different languages as being overly hung up on verifying objective (positive, reductionistic) facts about the world, separate from beliefs and practices associated with those facts. The latter (beliefs and practices), as mentioned above, are more important for the theories this book puts forward and more relevant to those interested in theories of geographical naming but with less desire to study word particles or Old English (Gelling, 1978).

Second, place-naming practices are implicated in the production of geo-NTNs, their effects, traces, and in the place names (references) themselves and their referents. Both reference and referent can be said to evolve in line with practice. Names change because beliefs and languages change, but also because the world itself sometimes changes—'from beneath' as it were. Worlds offer foundations in the form of features (e.g. hills, lakes, or geological strata) to which one is able to refer; the style of reference 'comes after.' Therefore, we posit foundations as real even though later (neogeographical) place-naming practices do not always assume them to be there in any positive sense. In this sense the neogeographical chapter on names (chapter 6) is post-foundational.

Third, this book seeks to avoid (it must be reiterated) the sometimes rigid determinism (albeit often of a weak variety) contained in Kripke's writings and theories, while retaining the rigor of Kripkean logic as it applies to (place) naming, its practices, networks, and effects. (Late) Wittgenstein and followers like Hanna and Harrison are the right people for the job. It is one thing to state this, and quite another for the present work to actually come to embody such a goal, that of determining the nature of geographical names in a non-rigid way while retaining the flavor and rigor of the Kripkean framework. Wittgenstein is brought in as an ally to this endeavor. So many disciplines have benefited from Wittgenstein's later work that it is worth geography joining the crowd on this. Because Wittgenstein also happened to focus so very much on both names and space (i.e. Wittgenstein often looked to geography or geographical metaphors to illuminate his concepts) in his later work, I also feel justified in doing so in this book. An added 'side' benefit, if you will, is that Wittgenstein's work is also so often focused on practices and beliefs that it is appropriate going the 'other way' as well (i.e. from geography to Wittgenstein), to look back from the perspective of geography to philosophy of language as it is embodied by its greatest practitioner.

The study of geographical names is exciting for a number of reasons. In landscapes, conversations, and on maps, names take on meaning well beyond their potential as mere labels. For labeling, in a physical geographical sense, would merely entail numerical sequencing or some other algorithm for the production of 'tagged' landscapes. The geographies of human lives are far richer. Not only do names 'spark' something in people, giving meaning far beyond their basic (but not sole) function as tags or labels; geographical names anchor identities in space in much the same way as person names anchor identities through time (Eades, 2015a). Corroboration of this claim, through the production of a geographical theoretical framework for examining the nature of geographical names through space and time, with evidence from place-name databases and primary texts, is the driving force behind this book.

As a geographical endeavor, toponymy has connotations both popular and academic. A popular view of geography might reduce the discipline to a trivial endeavor associated with memorization of lists (e.g. of capital cities). With some exceptions (e.g. Rose-Redwood et al., 2009; Rose-Redwood, 2008; Berg and Vuolteenaho, 2009) the contemporary situation in toponymic studies has not seen a great deal of improvement, with an oft-noted hiatus in toponymic inquiry among geographers largely continuing its inertia and lack of forward movement. The exceptions cited above give hope: studies from a perspective on power and politics of place-naming practices are beginning to appear, but movement is slow and halting at best. In reaction to, or often in spite of, the inertia around geographical naming in geography, we have in the discipline some 'outlier academics,' for instance Olsson (2007) and Pred (1990), who maintain an interest in names (and geography) in large part as an outcome of their philosophical sophistication (and eccentricity) of their thought. Indeed, Olsson demonstrates an interest in Kripke in *Abysmal*; while Pred has cited Wittgenstein extensively in his work.

But this book is concerned with much more than toponymy. 'Name' is often a name for something else: metaphors, discourses, practices, and logical 'atoms' of knowledge (McGinn, 2015; Wittgenstein, 2001; Frege, 1993). To say that names in space are metaphors for place-making practices and discourses is to say that names mean more than they seem to. Their utterance can even seem to be 'magical' at times, a theme that is explored below in relation to boundary-making rituals, practices, and systems of representation from medieval (but continuing into present) times (Mauss, 2001; Pounds, 2000).

There are senses (many in fact) in which names represent spatialized phenomena without constituting toponymies proper. Take taxonomy, for example. Foucault (2002) points out that spatialized classifying grids were a visual means used by geographers (and others) from roughly the 1700s onwards for reducing natural phenomena to their names (Gregory, 1994). Linnaean classification resulted in unique names (visualized as) corresponding to unique species. Later, in Darwin's view, named species evolved, transforming by natural (and other kinds of) selection along axes (now called cladograms) into new species. The grids (of names) thus became mobile, but no less visual for all that.

Geographical (or spatialized) names could also represent individual trajectories in space of mobile subjects, traceable and nameable through the use of maps, mapping devices, and metaphors. For example, 'tagged' Twitter accounts can include location in the form of coordinates (e.g. latitude and longitude) which are then easily mapped using computer software and programming languages such as Python, or sites that automatically visualize such 'geotagged' tweets as a service for users without the programming background to construct such platforms themselves (de Souza e Silva and Frith, 2012). 'Geotags' and 'tweets' are, for the purposes of this book, instances of geographical names.

In a book covering the geography of names, it will be important to have a solid working definition of the term *name*. Names are defined, for our purposes, in a Fregean sense as "singular noun-phrases that are used to refer to particular things" (Moore, 1993, page 1). Note that this definition does not contain reference to geography and that it is distinct from the idea of a 'proper' geographical name. To clarify, we must explore definitions of concepts related to names, including *naming practices, meaning, sense, reference, identity,* and *truth,* all of which are explored below and throughout the book.

Some context is necessary. One of the implications of defining names in the broadest possible way is that it short-circuits the possibility of a trivializing focus on toponymy or place names. We focus here more on *naming practices* as part of what it means to *do* geography in both lay and academic senses. Names form parts of word/world relationships that are essential to both human and physical geographical research. As such, practices are essential to both and meanings or senses (not to mention truth-conditions) of names that cannot simply be 'read off' directly from nature (Hanna and Harrison, 2004). In his or her practice, the geomorphologist follows rules of thumb (practices) at least as often as the ethnographically or qualitatively oriented human geographer. Words, in the form of names as defined above, mesh with worlds (human and physical) through naming practices that bridge the two in a two-step linkage of word-practice-world, with practices forming bridges between words and worlds. This book covers the relationship between words and worlds from a geographical perspective using a theory of practice (also known as rules) informed by the writings of Wittgenstein (2001 and 2009).

In *Philosophical Investigations*, Wittgenstein explored names in depth, often using geographical metaphors to do so. Wittgenstein's practice, then, was to use ideas about space to elucidate, explain, and analyze senses of the idea of naming practices, and then to give examples. Names, for Wittgenstein (2009, page 18), could be posted at certain addresses ("the post at which we station the word"), an analogy for assigning reference to words. If the assignment of reference (and by extension meaning) to names (made up of words) was, for

Wittgenstein, distinctly spatial, it was at the same time sometimes not spatial at all (or only quasi-spatial). Different kinds of words were also (metaphorically) tools or levers in a train locomotive, looking superficially very similar to each other (all ink marks on a page; all metal rods with plastic knobs) but in fact with very different functions (Wittgenstein, 2009, page 10). Words as parts of language are pieces in language games, analogous as well to chess pieces with different moves, rules, and strategies of use. Words, again, could be like pictures or slabs of information (Wittgenstein, 2009, page 12). In other words, Wittgenstein's practices for elucidating names are varied. The title of this book with its 'The' predicating geography, in a singular sense, is meant to convey precisely that names have geographies in the plural, but also that we are focusing (again, precisely) on the *geographical* in distinction to the pictorial, ostensive, ludic, or utilitarian senses of naming, though these may in fact enter the picture in various ways below so long as they have geographical senses. 'The' is meant, therefore, to specify a broader geography of names, not reduce this space to one way of thinking.

By the *meaning* of a geographical name we thus mean much more than the coordinates and linguistic translation of a toponym. We mean very precisely, by referring to the meaning of a name, that it fits into naming practices for producing relationships between words that form, or constitute, the name, and the worlds of practitioners whose practices (cultural, physical, or otherwise) link up with those words through uses in writing, in performance, and as utterances.

Routes and representations

Taylor (1985, page 252) has pointed out that Frege's sense of 'sense' is intimately tied to both the representation of a thing or an idea and the route by which we come to know that representation (which Moore (1992, page 2) calls mode of presentation) as essential to the sense of a name. The main point to keep in mind is that the same place may be referred to by two different names with different *senses* without affecting the *reference* of either name. Let's take 'London' and 'Londres' for example, refer to the same place. The utterance or writing of the word may be different for the French speaker, or it may be otherwise performed differently (such as by walking around the city) and in this way evoke a different sense of the name. The particular way in which the actual city is presented to the mind of that name's speaker, writer, or walker is its representation. So, while two names with different senses can have the same reference, the converse is that two names with exactly the same sense cannot have different references (Moore, 1993, page 2).

Taylor (1985, page 252) points out that

> [s]pecifying sense is specifying the speaker's/hearer's route into the reference. But this Fregean image of a route invokes the underlying activity. Words are not just attached to referents like correlations we meet in

nature; they are used to grasp those referents, that is, they figure in an activity.

The idea of an activity is analogous, for our purposes, to practices, as human activities for sustaining meanings of named aspects of the world. Words are constantly of the world in whatever form they take in human minds (internally) and as creations (externally) in the form of maps, images, paintings, scientific publications, poetry and the like.

Some geographical thinkers hint at alleged incompatibilities between practices and representations (Driver, 2001; Gregory, 1994) as though the two would not mix or somehow represent a logical paradox. This book, and my previous one (Eades, 2015a), have been driven in part by a need to break down dichotomies that would strictly divide practices from representations. A post-representational turn in geography (Kitchin and Dodge, 2007) is pushing the discipline, on the human side, away from representations as being falsely 'fixing' of geographical phenomena or as lying below the level of interest, as in the more-than-representational sub-genre, in a somewhat reactive and new so-called 'non-representational' geographical tradition (Thrift, 2007).

On the contrary, it would be hard to think of something more representational than a name. Names claim to stand in for the thing named. We tend to think of objects as essentially synonymous with their names, while also thinking of names as labels that reflect truthfully (or not as the case may be) 'what's in the tin.' At the same time, as demonstrated below, and as shown amply by Wittgenstein (2009), names are as essential to practices as they are to human *identity* through their use values in religious, cultural, and scientific practices and thus also human notions of *truth* and belief. Wittgenstein (2009) touches on these aspects and more, but it behoves us now to bring a more Kripkean slant to the discussion for the following reason. Kripke tested a divide between essentialist views of names that were on the one hand rigid designators (proper names), *a posteriori* referring to persons (Kripke refers to Nixon) and places (both Madagascar and Everest are two of Kripke's favorite examples). On the other hand, Kripke used the very anti-essentialist (and almost post-modern) idea of possible worlds to prove such necessities. The following discussion is inspired by Kripke's so-called puzzle about belief, as explored by Fitch (2004), and is intended to tease out some of the complexities of why exploring names as both descriptors and rigid designators is so important to geography and belief.

Steam governors and thermostats

To what do we refer when we refer to London and how do we know when we're there? Note that this two-part question contains aspects of a temporal and spatial nature. Another way of phrasing the question might be: where is London located exactly and how long does it take to get there (assuming we are not there already)? Both questions are almost impossible to answer quantitatively because geographical reference is *performed, distributed*, and *mobile*. Take the *National Geographic Atlas of the World* (or a similar world atlas) as a starting point, opening (in the reader's imagination if a copy is not conveniently to hand) to the page showing the United Kingdom. A child interested in geography might say, "I want to go to the capital of the United Kingdom" (after his or her parents explain what a capital is and indicate the areas referred to by the words 'the United Kingdom'). They could then easily refer to London on the map. In this formulation, 'the capital of the United Kingdom' and 'London' are synonyms and, as such, have the same sense and reference. The act of pointing to London on the map is a performance that requires background knowledge of map reading. Given the scale of a map of the UK in the *National Geographic Atlas*, it is very easy to point to London without ambiguity.

Now, for the sake of argument, let's say that the family in question decides to visit London from where they live in Scotland. How they decide to go to London will have an effect on when they decide they have arrived, whether by train, plane, or automobile. It can be stated as: "We will know we're in London when x" (or in the past tense, "We knew we were in London when x"). In this instance, x can take a variety of values, and is not fixed to any one feature of London (e.g. the 'Big Ben' clock on the parliament buildings); x can also with equal validity be posited as any of the following:

- we see the Shard (the tallest building in London)
- we see Big Ben (the most photographed object in London)
- we cross/get onto the M25 (motorway that encircles London)
- we're at Heathrow airport
- we are at King's Cross Station
- we can see any or all (or a combination) of the buildings going by the names of: the Shard, the Gherkin, the Cheese Grater, and the Walkie-Talkie (to use the vernacular)
- we can see 'London' from above (from a plane or tall building)
- we can see the Thames River.

The referent of London is thus dependent on the mode of transportation, and in this sense it is mobile; it is also dependent on whether one thinks of London as a discrete, binary thing (you're in or out, you're there or not, with no in-between) or as continuous, with no real edge or centre point. Whichever way one looks at it, however, London is a *distributed* phenomenon. To the extent that there really is no one right answer to where London is located, how one answers the question of when is one in London depends in large part upon whether one is persuaded by a *steam governor* model of geographical reference or, on the contrary, one buys more into a *thermostat* model. This will have a significant effect upon one's certainty with respect to the ability to know when one is 'there' (i.e. in London) (Dietrich and Markman, 2003). Steam governors and thermostats clearly need a bit more explaining, and we turn to this now, starting with the latter.

A thermostat model of geographical reference uses an analogy of the device containing a bimetallic strip for regulating heat. When the strip cools or heats it curls due to differential contraction of the two metals it contains, one on either side of the strip, placed back to back. The switch is turned either on or off depending on the direction of the curl (either towards or away from the contact which completes a circuit), which in turn activates or deactivates a machine that heats the room. The point is that the thermocouple can exist only in one of two states (on/off) and it is in this sense binary.

The analogy for geographic reference is that one is either in a (named) place or not, with no in-between. The case of Royal Holloway (part of the University of London) is interesting in this regard (and especially so since the author works there as a lecturer). It is clearly outside the M25, the furthest limit for many defining London (i.e. if you're outside the M25 orbital loop road you're not in London), but Royal Holloway is also clearly part of the University of London and, as such, *in* London. Does it matter that it is also located only just outside the M25 (i.e. almost literally a stone's throw from it) and, as it happens, in the small suburban/rural village of Egham, in the county of Surrey? According to the thermostat model you're either in London or not, so for us to be both in London and at the Royal Holloway campus one would have to add to our list of possible x values something like:

- we can see the Founder's Building (iconic at Royal Holloway)

for us to count as being in London (not to mention Egham, Surrey) according to the thermostat model. It is worth noting that one can see the Founder's Building from the M25 motorway and therefore from within the main part of London (defined as that circled by the M25).

Thermostat model believers would most likely accept the following 'binary' situations (imagining a 'click' as the connection is made as analogous to the following occurring) as criteria for being in London:

- crossing the M25 in the direction of the city centre
- Shard becomes visible
- Big Ben becomes visible
- touch down at Heathrow
- entering King's Cross Station.

The steam governor model of geographical reference, on the other hand, uses the analogy of the steam locomotive, in which feedback from what is called a governor increases or decreases the amount of power being delivered to the train's drive depending upon the state of the system. The mechanics are not important, but the idea of a continuum and the contrast to the thermo-couple are important for the analogy to work (Dietrich and Markman, 2003). If we step outside the binary paradigm of knowing when we are either there (in London) or not, London becomes a continuum of effects, some more

London-y, some less so. Increasing traffic volume combined with proximity to the M25 or the ability to see Wembley Stadium (but not The Shard due to haze) combined with being at Royal Holloway are indicators, at least, that one is close to London (if not 'in' it already). One then notices (if traveling by car, train, or plane) that the Thames River seems larger (wider) and is traversed several times, and suddenly (perhaps at the same moment that one sees Big Ben) one knows that the time has arrived when one can with certainty say, "We're there, we're in London."

Frege refers to when/where questions like this as being conditional clauses. The presence or visibility of a feature is a condition for knowing something (i.e. that one is 'there'). When/where questions can be distinguished from who/what questions in the sense that the objects of reference for the latter are more concrete (more easily bounded) (Frege, 1993, page 38). In any case, it is the 'thing itself' that is named, and in this sense we maintain a critical realism with respect to the 'proper' referent of the objects (persons or places) in question.

Both place and person names have senses of being 'proper' names, but as has been made clear in the above discussion of thermostats and steam governors, they (places and persons) are far from being similar kinds of thing. While, for both, descriptions may not stand in for the names, proper, in the case of places the object referred to by a place-name is much more distributed, indefinite, and ambiguous than is the case for person names. What Frege refers to as the "cognitive value" (Frege, 1993, page 42) of names means that descriptions of features and the names themselves have both steam governor and thermostat aspects, but the idea of names as discrete, referring things tends to sway cognition in the direction of the thermostat model (Eades, 2015a). Because of this, it is the favored *mode of operation* for the present work.

Rituals, representations, and rogations

The primary question concerning this book is: *what is the nature of the geographical name, and how (and where) has it changed through time?* I posit that names and naming practices form intricate parts of not only representations (as explored above) but also rituals and rogations and that they are inherently political (and contested) in nature. I present evidence for these claims gathered from both historical and contemporary archives, publications, pamphlets, and databases. Indigenous place names are explored here and in earlier work (Eades, 2015a) by examining north and northwest Canadian indigenous databases; publications such as Boas' (1934) *Geographical Names of the Kwakiutl Indians* and another by the Smithsonian Institute on Mayan place glyphs (Stuart, 1994). These databases and works provide a foundation for exploring geographical names that is at once descriptivist, ostensive, and intergenerational in the sense that indigenous spatial identities and learning take place in interaction both with the land and with elders and peers while participating (seeing and doing) in journeys that include hunting and fishing activities (Eades, 2015a). This takes us well beyond descriptive lists of place names held in institutes and archives, into performative mappings that, in indigenous life-worlds, are more about seeing and doing (ostensive learning) while outdoors with one's family than about dusty archival research.

During such activities on the land, the representation of surrounding space—for the indigenous traveler (chapter 2) and religious devotee (chapter 3) alike (Jones, 1954)—was implicated in a range of demonstrative and ritual activities that include naming. 'Baptismal' (or original) moments of geographical naming in indigenous and religious contexts often seem to imply that features in the landscape can be seen unproblematically to exist prior to the assignment of linguistic references to them, but also that the names reflect 'true' aspects of the named places (referents in the real world). The sense of what 'ground truth' entails in relation to evolving sets of naming practices is part of what this book explores. Chapter 4 considers how indigenous and religious naming practices lead to connotations not so much of being 'indigenous' but of something internal to the human body and brain. Chapter 5 explores the politics of place names and their uses in counter-mapping, a method of mapping against the (cartographic) power of the state, often exploited by indigenous groups in Canada. Chapter 6 moves into new geographies, aka neogeographies, of naming, while chapter 7 concludes by examining the implications of all that follows in the main chapters.

There is another way in which naming, especially for the First Nations peoples of Canada, is important. At the time of this writing, the Truth and Reconciliation Commission (TRC) initial recommendations have just been released in Canada. An early version of this chapter (Eades, 2015b), entitled "Geographical Naming and Necessity," was presented at the Canadian Association of Geographers annual meeting in Vancouver in June 2015. An audience member (an Anishinaabe man) asked about the relationship between the TRC testimonials and geographical naming. My answer seemed to me obvious (and perhaps thus insufficient) at the time, but in retrospect I find it surprising that I had not noticed the parallel before. I believe that TRC processes of naming, in the sense of describing what went on in the Indian residential schools of the late nineteenth and early twentieth centuries, operates in parallel to indigenous and First Nations efforts to reclaim identities lost, erased, or destroyed through what Canada is not prepared (because of the TRC) to describe as cultural genocide (Niezen, 2013).

With its partial focus on Canada, this book attempts to address a sub-question beyond any to do with the nature of geographical names. This sub-question can be approached via a flawed observation on Kripke made by McGinn (2015). Kripke (1981, page 81) states that "we know Cicero was the man who first denounced Catiline. Well, that's good. That really picks someone out uniquely. However, there is a problem, because this description contains another name, namely 'Catiline'." Kripke is trying to avoid the circularity condition (C) spelled out in Lecture II of *Naming and Necessity*, delivered on January 22, 1970 at Princeton University.

The circularity condition states that "for any successful theory, the account must not be circular. The properties which are used in the vote must not themselves involve the notion of reference in such a way that it is ultimately impossible to eliminate" (Kripke, 1981, page 71). By vote Kripke means a set of properties that uniquely pick out some object (person, place, or thing). McGinn (2015, page 53) makes an interesting statement, the significance of which will become apparent shortly, that is directly refuted by the two Kripke quotes above. McGinn claims that Kripke makes no mention of impure descriptions or those that define a name in terms of other names, despite what Kripke wrote on page 71 of *Naming and Necessity*.

McGinn writes that the name 'Aristotle' can be replaced by the definite description 'the best pupil of Plato.' To quote McGinn (2015, page 53),

> notice that this description contains a name, 'Plato.' Many of these uniquely identifying descriptions contain such names. But according to the description theory, all names are equivalent to descriptions. What then is meant by the name 'Plato'? The name 'Plato' cannot abbreviate the definite description "the teacher of Aristotle" because that definition would be circular. To refer to Plato, we must create a new definite description. We could say, 'the most famous philosopher of ancient Greece,' but then the question would arise as to what the name 'Greece' means. The point is that the uniquely identifying definite descriptions themselves contain another name. To explain what that name means, the descriptions continue to regress to descriptions containing other names. This issue raises serious problems for the description theory, since names are supposed to depend ultimately on descriptions for their reference.

Definite descriptions are Fregean constructs that Kripke set out to refute. According to scholarly opinion (Burgess, 2013; Berger, 2011; Fitch, 2004), with the possible exception of McGinn (2015) himself, Kripke was successful in this refutation, using the idea of possible worlds to spell out how names retain referents that in other worlds have radically different properties. What if Aristotle had decided to study music, or been born with a brain defect? He would still have been Aristotle and the name would still have referred to him uniquely. Analogously, what does 'Greece' refer to? Historically speaking, could this referent have turned out differently? In what ways? We think of many things (both ancient and modern) when we utter the name Greece, but it does not refer in any straightforward way. Think of the Elgin Marbles. To whom would Greece have us return these spoils of colonial endeavor? We refer to Greece and other geographical entities through time from a baptismal moment when the name first occurs to the present day. The form and content of the name may change, as might its referent, and there are many variations in how names evolve between origin and present usage. What matters is that the names continue to refer and that they do so through a historical chain of communication (Kripke, 1981, pages 91–97). The word "Celt," coined by

Caesar—in the course of colonial conquest again—is but another example of a coinage that has come to have different connotations through time and a variety of areal extents and identities resulting from a coinage that originally applied to what is now known as Brittany, in the northwest of France.

The question for audiences of *The Geography of Names* is how all this applies to theories of geographical reference. "Geographical Naming and Necessity" was presented at the Canadian Association of Geographers (CAG) meeting on 2 June, 2015 in Vancouver, British Columbia. It presented findings associated with reading from Kripke (1981), Wittgenstein (2009), and Hanna and Harrison (2004), applying insights from these three key texts to geographical naming systems, with necessary and sufficient referring conditions and special reference to the 'problem' of indigenous names that are the subject of chapter 2. Does the question of ancient names pertain in Canada? Who renounced whom, and is there a place for indigeneity in all this? How do we politicize the question in light of First Nations struggles to maintain the integrity of the land and fight damaging legacies of residential schooling systems, in courts and with each other and themselves? Where, indeed, do indigenous senses of self, identity, naming, and land fit into (primarily) white man's theory?

At the end of my presentation at the CAG a participant noted that its content coincided with the TRC findings, and a prescient question was asked about the fit of my findings with those recommendations. My answer alluded to parallel trajectories and healing both on the land and in everyday life. I believe that examination of all aspects of names, naming practices, and systems and a blurring of lines between past and present, indigenous and non-indigenous, ancient and modern is the most productive approach for the purposes of this book and for beginning to answer its primary and sub-questions. The sub-question, that of circularity, communication, and renunciation, is of primary importance to the healing process many First Nations and indigenous individuals and communities face today. Experiences shared in the TRC disallowed naming of abusers in the final publications produced by the commissioners, allowing for full disclosure in the oral testimony without allowing blame to be inscribed with specific finality (Niezen, 2013). As mentioned to the questioner at the CAG, the parallel trajectory of naming geographical names to do with the land allows another very important aspect of healing to occur, before and after the sharing of traumatic memories of residential school experiences with the TRC.

The other question asked of me at the CAG (by a young woman) was about dead white man's theory and I addressed it by saying that I believe in building bridges between indigenous and non-indigenous research. A particular strength of doing research on and in Canada is that these bridges are already being built; they are in various stages of construction (Davis, 2010). I asked myself, however, whether this was good enough. After all, dead (and, for that matter, many live) white men formed a core component of those being denounced or simply named by residential school survivors abused by those (mostly) men who remained nameless after their identification in the official archive. Naming my influences (dead and alive), it is noteworthy that while many of the citations in this book are to classical or European (white) thinkers, including Wittgenstein and Kripke, most are still alive and many are women. One of the most important theoretical texts included in the framework for examining our main research questions here is Hanna and Harrison's *Word and World* (2004), an innovative and uniquely insightful text in the literature on philosophy of language, one of the few authored by a woman and one of the fewer still that make reference to hunter-gatherer societies and geographical naming.

Chapter 2 also includes a good deal of framing using the works of Turner (2014), a female ethno-botanist who has devoted her life to examining the plants of northwestern North America from ancient to modern times. In a way, this is a white man's telling, as I make use of texts written by voices speaking from the margins in combination with texts by those who definitely do not (i.e. Kripke and Wittgenstein). But as I mentioned to the young woman at the CAG, I am not going to give up those texts any time soon. I remain agnostic as to the positionality of the creator of a tool, valuing the worth of a tool by its usefulness. Turner's (2014) tools remain at least as useful as those of Hanna and Harrison (2004), who in turn rely upon Sapir (1958) and Wittgenstein (2009), respectively. As Cameron (2015) has pointed out, it is time that white people begin to refine their narratives, to change the way they speak about indigenous peoples, and to take responsibility for the shape and content of the resulting stories. I take Cameron's challenge very seriously. The results are for the reader alone to judge.

References

Aunger, Robert. 2002. *The Electric Meme: A New Theory of How We Think*. New York: The Free Press.
Berg, Lawrence and Vuolteenaho, Jani (eds). 2009. *Critical Toponymies: The Contested Politics of Place-Naming*. Farnham: Ashgate.
Berger, Alan (ed.). 2011. *Saul Kripke*. Cambridge: Cambridge University Press.
Berger, Alan. 2002. *Terms and Truth: Reference Direct and Anaphoric*. Cambridge, MA: MIT Press.
Blackmore, Susan. 2000. *The Meme Machine*. Oxford: Oxford University Press.
Boas, Franz. 1934. *Geographical Names of the Kwakiutl Indians*, Columbia University Contributions to Anthropology, vol. 20. Columbia: Columbia University Press.
Boyd, Robert and Richerson, Peter. 1985. *Culture and the Evolutionary Process*. Chicago: University of Chicago Press.
Burgess, John P. 2013. *Kripke*. Cambridge and Malden: Polity.
Cameron, Emilie. 2015. *Far Off Metal River: Inuit Lands, Settler Stories, and the Making of the Contemporary Arctic*. Vancouver: UBC Press.
Cavalli-Sforza, Luigi-Luca and Feldman, Marcus. 1981. *Cultural Transmission and Evolution: A Quantitative Approach*. Princeton: Princeton University Press.
Davis, Lynne (ed.). 2010. *Alliances: Re/Envisioning Indigenous-Non-Indigenous Relationships*. Toronto: University of Toronto Press.

Dawkins, Richard. 1976. *The Selfish Gene.* Oxford: Oxford University Press.

Dawkins, Richard. 1982. *The Extended Phenotype.* Oxford: Oxford University Press.

De Souza e Silva, Adriana and Frith, Jordan. 2012. *Mobile Interfaces in Public Spaces: Locational Privacy, Control, and Urban Sociability.* New York and Abingdon: Routledge/Taylor & Francis.

Dietrich, Eric and Markman, Arthur. 2003. Discrete Thoughts: Why Cognition Must Use Discrete Representations. *Mind & Language.* 18(1). 95–119.

Distin, Kate. 2005. *The Selfish Meme.* Cambridge: Cambridge University Press.

Driver, Felix. 2001. *Geography Militant: Cultures of Exploration and Empire.* Oxford and Malden: Blackwell.

Eades, Gwilym. 2015a. *Maps and Memes: Redrawing Culture, Place, and Identity in Indigenous Communities.* Montreal and Kingston: McGill-Queen's University Press.

Eades, Gwilym. 2015b. Geographical Naming and Necessity. Canadian Association of Geographers Annual Meeting, Vancouver, British Columbia.

Fitch, G.W. 2004. *Saul Kripke.* Chesham: Acumen.

Foucault, Michel. 2002. *The Order of Things.* Abingdon and New York: Routledge Classics.

Frege, Gottlob. 1993. On Sense and Reference, in Moore, A.W. (ed.) *Meaning and Reference.* Oxford: Oxford University Press.

Gelling, Margaret. 1978. *Signposts to the Past.* Chichester: Phillimore.

Gregory, Derek. 1994. *Geographical Imaginations.* Cambridge, MA and Oxford: Basil Blackwell.

Hanna, Patricia and Harrison, Bernard. 2004. *Word and World: Practice and the Foundations of Language.* Cambridge: Cambridge University Press.

Hawthorn, Tom. 2012. Prince Rupert Visionary Hauled From The Titanic's Wreck. *The Globe and Mail.* 11 April. Accessed January 19, 2016.

Jablonka, Eva and Lamb, Marion J. 2005. *Evolution in Four Dimensions: Genetic, Epigenetic, Behavioral, and Symbolic Variation in the History of Life.* Cambridge, MA: MIT Press.

Jones, Francis. 1954. *The Holy Wells of Wales.* Cardiff: University of Wales Press.

Kitchin, Rob and Dodge, Martin. 2007. Rethinking Maps. *Progress in Human Geography.* 31(3). 331–344.

Kripke, Saul. 1981. *Naming and Necessity.* Malden: Blackwell.

Kripke, Saul. 2011. A Puzzle About Belief, in *Philosophical Troubles: Collected Papers*, vol. 1. Oxford: Oxford University Press. 125–161.

Lewens, Tim. 2015. *Cultural Evolution: Conceptual Challenges.* Oxford: Oxford University Press.

Mauss, Marcel. 2001. *A General Theory of Magic.* Abingdon and New York: Routledge Classics.

McGinn, Colin. 2015. *Philosophy of Language: The Classics Explained.* Cambridge: MIT Press.

Monmonier, Mark. 2006. *From Squaw Tit to Whorehouse Meadow: How Maps Name, Claim, and Inflame.* Chicago: University of Chicago Press.

Moore, A.W. 1993. Introduction, in Moore, A.W. (ed.) *Meaning and Reference.* Oxford: Oxford University Press.

Niezen, Ronald. 2013. *Truth and Indignation: Canada's Truth and Reconciliation Commission on Indian Residential Schools.* Toronto: University of Toronto Press.

Olsson, Gunnar. 2007. *Abysmal: A Critique of Cartographic Reason.* Chicago: University of Chicago Press.

Plato. 2008. *Timaeus and Critias*. Oxford: Oxford University Press.

Pounds, N.J.G. 2000. *A History of the English Parish*. Cambridge: Cambridge University Press.

Pred, Allan. 1990. *Lost Words and Lost Worlds: Modernity and the Language of Everyday Life in Late Nineteenth-Century Stockholm*. Cambridge: Cambridge University Press.

Richerson, Peter and Boyd, Robert. 2005. *Not By Genes Alone: How Culture Transformed Human Evolution*. Chicago: University of Chicago Press.

Rose-Redwood, Reuben. 2008. From Number to Name: Symbolic Capital, Places of Memory, and the Politics of Street Renaming in New York City. *Social and Cultural Geography*. 9(4). 431–452.

Rose-Redwood, Reuben, Alderman, Derek and Azaryahu, Maoz. 2009. Geographies of Cartographic Inscription: New Directions in Critical Place-Name Studies. *Progress in Human Geography*. 34(4). 453–470.

Runciman, W.G. 2009. *The Theory of Cultural and Social Selection*. Cambridge: Cambridge University Press.

Sapir, Edward. 1958. Time Perspective in Aboriginal American Culture: A Study in Method, in Mandelbaum, David (ed.) *Selected Writings of Edward Sapir*. Berkeley and Los Angeles: University of California Press.

Stuart, David. 1994. *Classic Maya Place Names*. Washington, DC: Smithsonian.

Taylor, Charles. 1985. *Human Agency and Language*. Cambridge: Cambridge University Press.

Thrift, Nigel. 2007. *Non-Representational Theory: Space, Politics, Affect*. Abingdon: Routledge.

Turner, Nancy. 2014. *Ancient Pathways, Ancestral Knowledge*. Montreal and Kingston: McGill-Queen's University Press.

Wittgenstein, Ludwig. 2001. *Tractatus Logico-Philosophicus*. Abingdon and New York: Routledge Classics.

Wittgenstein, Ludwig. 2009. *Philosophical Investigations*, 4th ed. Chichester: Wiley-Blackwell.

2 Indigeneity and geographical naming

Landscape, language, and environment

Sapir (1958, pages 436–437) notes the importance of place names in ethno-graphic study but elaborates very little. What he does say, however, counts for a lot and his words are worth quoting at the start of this chapter in order to set the stage for what follows. Sapir's section on place names is short but embedded in a larger discussion on methods for exploring the diffusion of elements of material culture in landscape. Place names are noted as being extremely important to the establishment of certain facts of habitation, sequencing, and extent of North American aboriginal populations. The ability to interpret meanings of specific place names is tied, according to Sapir, to the amount of time elapsed since a group has come into contact with a landscape feature. For example,

> Mt. Shasta, in northern California, is visible to a considerable number of distinct tribes. The Hupa call it … a descriptive term meaning "white mountain"; while the Yana have a distinctive term for it, wa'galu, which does not yield to analysis. We may infer from this that the Hupa, as an Athapaskan-speaking tribe, are newcomers in northern California as compared with the Yana, a conclusion that is certainly corroborated by other evidence.

Sapir's observation forms a baseline for a geography interested in place names beyond bare meanings, facts, and coordinates. What the above obser-vation begins to get at is not mere chronology but the fact that things could have been otherwise in the landscape, that the shape of a geography resulting from human interactions over thousands of years in North America, pre-contact, was not set in stone, and that ethnographic boundaries could have ended up looking radically different from that shown on contemporary maps. These would note the presence of Mount Shasta as a central feature, a high ground with spiritual and physical significance, one that by its very nature serves to define both a watershed and a moment in ethnographic time. The Hupa and the Yana define themselves, not always consciously, in relation to each other. The Huna named what we call Mount Shasta much later in time. What are the implications of this later naming for relations between neighbors?

Names are political and contested, and they are caught up in scientific methods and nomenclatures imposed from outside by explorers, academics, and other newcomers. Hupa and Yana are themselves names, given by peoples to themselves with territorial extents defined relationally through time, prior to outside contact. Thus, if we continue to use the name Mount Shasta at the same time as we use both Hupa and Yana we find ourselves in a conundrum, caught up in possible worlds thinking. Could it have been otherwise? What of a world where 'Mount Shasta' (as a name, or even as a physical entity) ceased to exist? Could a revisionist line of thinking take us back to origins, unsmudged by colonial incursion and subsequent renaming of entire indigenous life-worlds? These kinds of scenarios, some science fiction, others pure fantasy, will not be seriously entertained here.

Wikipedia tells us that Mount Shasta is an inactive volcano and the fifth highest peak in California. What of this fact? For local and indigenous populations, speaking historically, it is almost certain to be of little importance due to lack of relevance to everyday life. Naming, from native perspectives, was not focused on 'highest,' hierarchy, or the application of person names to landscape features. Instead, things on the mountain itself would have been named, in densities according to their importance to livelihood (making a living), spirituality, and wayfinding imperatives, not to mention possible mythical aspects explaining the creation of the landscape—indeed the whole of the world from time immemorial. It would appear from Sapir's observations above that since the name 'Shasta' does not, in fact, match the indigenous names for the feature lying at the boundaries of Hupa and Yana territories, it could be something of a misnomer or even a transference from another feature to the peak we currently associate with the name. Such is the way, as we will see below, and often the folly of outside attempts to impose names on features long known to local inhabitants.

Thornton (2008), Mark et al. (2011) and especially Turner (2014) have noted the importance of densities, perspectives, and ecologies of indigenous naming systems, and how they differ from non-indigenous toponymies. In northwestern North America in contemporary BC and Alaska, for example, glacial refuges on coasts and mountaintops are tied to the spread of plant species important for indigenous peoples in the area. Human settlements were tied to so-called 'edge' areas such as shorelines, treelines, and river terraces, and indigenous place-naming patterns can be expected to vary according to these spatial constraints (Turner, 2014, page 77). More than simple tagging of areas of human settlement and activity, however, place names have been demonstrated to be part of indigenous (in the senses of internal as well as 'first') human databases for keeping track of large amounts of ecological information (Collignon, 2006; Johnson and Hunn, 2010) of relevance to everyday life, livelihood, and being on the land (and in place).

This explains, for example, why the highest part of a mountain, the point corresponding to its peak, is of less importance than the possibilities, indicated by place names, for harvesting edible or otherwise useful species of plant near treelines or at the edge of water. The place names themselves act like bridges or place-holders for the possibilities of what can be found across landscapes, combining ecological information with an indication of which species may be present or absent into names meant to evoke or remind the traveler or seeker of sustenance of what is there, at least potentially, in a specific named place. Non-indigenous place-naming systems contain no such information, or do so very differently. We might remember the name of a location of a rest stop on a highway with a particularly good restaurant, but it is unlikely to evoke the quality of the food in any way close to how indigenous place names have done so for millennia, in close contact with the earth, its ecology, and its other-than-human inhabitants.

Place names are tools, in Wittgenstein's (2009, page 15) sense, for making sense of the environment and earning a living, especially for indigenous hunting and gathering peoples and those whose livelihoods come from the land. This will become clearer below, and we will see that there are possible worlds (Kripke, 1981) to thrive in open to those whose knowledge of place is named in ways that bridge various kinds of knowledge across boundaries. Worlds of possibility opened by indigenous place-based knowledge systems indicate that local and traditional knowledge systems thrive and expand where ecologies are known intimately, but in many cases this tradition has broken down. We are coming to know, for example, how lack of engagement in post-modern (and even post-indigenous) contexts is damaging to environments due to impoverished relations with what it is all too easy to characterize as externalities, or those things that do not easily fit global economic models of progress. Words are tied, in indigenous life-worlds, to practices, and practices are, in turn, tied to worlds (Hanna and Harrison, 2004). It is to these worlds that we now turn.

Ecotopes and toponyms

Through his work on teleology and design in ancient Greek thought, Glacken (1967) has pointed out relationships between plant life as a basis for human life and naming. Strabo, we are told, points out in his *Geography* that, "in the descriptions of peoples bordering on the Red Sea and the Indian Ocean and the interior lands of Ethiopia," early inhabitants of the earth were named according to their diet: "the rhizophagi (root eaters), the hyolophagi (wood eaters), the spermatophagi (the seed eaters), and so on" (Glacken, 1967, page 21). In the same vein, it is later pointed out that the gods of the Mediterranean region were named with an eye to "personification of natural processes" that have a profound effect upon crops' success or failure (Glacken, 1967, page 36). Hence, naming was intensely geographical, as weather and climate vary both temporally and spatially, and the named gods could thus be seen as mobile and distinctly uneven or bounded in the dispersal of favorable conditions for crop success.

Glacken points out that for many of the ancients (of the Mediterranean), plant life was the basis for everything and that the purpose of life was directed towards a goal, that of the fulfillment of the higher callings of humankind. Thus, "there was a fullness and richness of life in nature; and plants existed for animals and animals for men" (Glacken, 1967, page 49). Living within a hierarchical society, men [sic] named the things around them as though they were gods (and the gods were above them bringing things into being with purpose and an eye always toward ultimate ends). We thus have a sense of necessity in the naming of the things of the earth, to keep track of its bounty and to keep things in their place. Democritus is famous for turning teleology on its head, for gutting the universe of purpose, but he turned to names no less than others of his time (e.g. Plato and Aristotle).

> 'Necessity' is the cause of atomic motion, the term signifying that everything is in motion for a reason … . Our world is not unique—it is only one of an infinite number. The concept of plurality of worlds is very clearly stated by Leucippus; … he is probably the originator of the atomist theory.
>
> (Glacken, 1967, page 64).

Philosophies of names (of particles) and (possible) worlds are thus not new but reach back to prehistoric times before Democritus and Leucippus and forward to philosophies of language positing names as necessary, *a posteriori*, for tracking objects through time (Kripke, 1981) via chains of communication. These in turn are posited to have originated in 'baptismal' moments (for people) or in ad hoc descriptions that 'stuck' at some point in time, through repetitive usage, and subsequent transmission and adaptation for purpose by future generations (Hough, 2015).

Indeed, names themselves can be seen as atoms of language, the most basic and necessary ways of keeping track of worlds in motion. Geographical names can be broken down into linguistic parts as Gelling and Cole (2000) and the philological approach to place names have amply demonstrated. Turner (2014) explores plant names extensively, noting various ways of tracking temporality and spread of names, their uses, and referents (not to mention uses of referents, or plant uses), in part breaking them down into sub-name components. These components could be seen almost like subatomic particles for tracking the spread of objects, peoples, and plant–person relationships and practices. Thus, in northwestern North America, the ancient indigenous inhabitants engaged with naming practices no less profoundly than the ancients of the Mediterranean, producing rich relationships through thousands of years of use and refinement of local plant and animal species.

Intergenerational ethno-ecological knowledge was intimately tied to place, instilled with meaning, metonymies, and rhythms of naming. Turner (2014, page 131) shows how northwest coast indigenous groups named plants based upon their relationships with animals, the latter category being one that implicitly included humans. Names also reflected use and descriptive appearance, infusing the places in which they were present with both inherent meaning and use in sets of interlinked practices. For example, mushrooms (of all varieties) are referred to by Gitksan and Haida as bird's or owl's hats (respectively, *gayda ts'uuts* and *st'aw dajaangaa*). Horsetails (*Equisetum*) are called goose food by the Dakelh people. Trembling aspen (*Populus tremuloides*) are known as 'dances around tree' (*yayaw'al'as*) amongst the Heiltsuk for the way the leaves of the aspen shift in the wind (Turner, 2014, pages 128–133). Visual aspects of relationships between plants have implications for livelihoods in indigenous life-worlds and are part of a poetics of naming and ethno-ecology, not simply aesthetic observations. Rather, the aesthetic and poetic elements of names play a key role in not just keeping people alive but enabling them to thrive in both traditional and modern landscapes (Hawkins and Straughan, 2015).

With such a variety of plants and animals, toponyms were essential for keeping track of all the names and the resources to which they refer. Rather than knowing exactly which species were available at each and every place, rules of thumb (practices) dictate that one could come very close to having *a priori* knowledge of exactly what resource would be where by knowing the names of places (Johnson and Hunn, 2010). Toponyms are thus part of 'human databases' or 'living geographic information systems' that are constrained, in much the same way as (non-human, machine) databases are constrained by their physical nature by being less than infinite in size. It has been posited that a human being can keep track of roughly (on average) 500 place names (Hunn, 1994) and that these place names act as parts of 'bridge entities' for storing much larger assemblages of ecological information (Hunn 1996) made sense of through the use of the concept of the ecotope (Eades, 2014; Hunn and Meilleur, 2010, pages 18–19).

An ecotope is defined as being roughly synonymous to

> 'kind of land,' 'biotope,' or 'habitat,' but we prefer ecotope because it does not imply a focus on land forms (versus feature of rivers, lakes, or the sea) nor on biological or, more often, botanical markers as definitive. Nor does the term ecotope have the ecological implications of the term 'habitat,' that is, a home for some particular species of plant or animal, including homo sapiens.
>
> (Hunn and Meilleur, 2010, page 16).

On the one hand, human individuals and groups have need of a system for keeping track of (named) places, which our brains are limited in ability to do. On the other hand, we have limited ability to keep track of more than a few hundred kinds of things we might be interested in as sources of livelihood (i.e. names of plants or animals). The ecotope performs a flexible bridging function for reducing a potentially very large amount of information (e.g. 500 places each holding potentially 500 plant and 500 animal species, or 500,000 pieces of information) into something manageable (e.g. 500 ecotopes that act as placeholders for a database of information).

Particularization

Levi-Strauss (1966 and 1983) has explored ways in which such particularizations of nature form parts of larger structures for making sense of indigenous life-worlds. More than that, they form scaffoldings for myths, granting larger meanings to particular worlds and, indeed, possible worlds like those described in the quantified philosophical language of Kripke (1981). Plants, animals, and landforms abound in *The Raw and the Cooked*, and one can see the outlines of ecotopic information systems in Levi-Straussian diagrammatics and structural transformations of this masterful (if sometimes bewildering) text. Levi-Strauss tackles names more directly in *The Savage Mind*, with clear connections between totemic species, clan names, and spatiality (or forms that spaces take) (Levi-Strauss, 1966, pages 172–186) hinting at names that are 'not yet properly proper': i.e. those that indicate through names without being addressed to individuals, referring instead to clans and other spatialized phenomena.

The proper name, on the other hand, is addressed to an individual and, thus, it is constituted in and through its ability to particularize (uniquely name) some-one. Another way of stating the same thing is to say that a person, place, or thing can (ultimately) be located (or not in the case of loss of a person, place, or thing) in time and space based upon the use of its name (Wittgenstein, 2009, page 15). Name-tracking networks (NTNs) make use of inscriptions (e.g. maps, wills, estate documents, or books), oral narratives, and other per-formances (such as name utterance in conversation or testimony) to establish continuity of the proper name through time (Hanna and Harrison, 2004). Every name thus has a history that is contingent but, paradoxically, defines it rigidly (Burgess, 2013). Precisely how this works—and what exactly is the mechanism at play in the tracking of names (and other cultural objects) across space and through time—is the subject of extensive debate in geo-graphy, anthropology, and philosophy. Evolutionary theorists have made the most interesting and detailed headway towards providing a satisfactory explanation for how culture (of which names are part) evolves, as described in the next section.

My aim in this book (and in Eades, 2015) is not to settle the matter once and for all but to provide an outline of practice for making sense of geographical names. As the title of the book suggests, the metaphors, analogies, and tools for doing so will be primarily spatial in nature, despite the temporal nature of much of the material. The reason for this apparent dichotomy lies in the inherently spatial nature of indigenous life-worlds and naming systems which are the foundation of this book (Nash, 2013; Mark et al., 2011). Much of naming itself is spatial and is hard-wired into the human body and its

performances of named places. Inscribed forms written into documents produced spatially (on the page), graphically (through drawings, gestures, and
trajectories), and otherwise with distinct x and y coordinates in mind (Evans,
1982, pages 151–170; O'Keefe and Nadel, 1978).

In other words, brains frame and focus attention in part through the use of
representations, or cognitive maps, that contain 'labels' or 'tags' for keeping
track of the locations of objects in the spaces the maps depict. This is a
contentious claim. Many would do away with representations either in part
(Rowlands, 2010) or entirely (Malafouris, 2013). Ismael (2007) offers a
balanced view of representations most compatible with the philosophy (and
the reality it ultimately claims to represent) espoused in this book and most
consistent with its aims. That is, representations (of named persons, places, or
things) exist both in minds and in the world, and there is no clear delineation
between the two other than to say they both exist and that, sometimes, named
things (and representations thereof) may span brains, bodies, and extended space.
This way of seeing things is in keeping with new theories of extended mind
(Clark, 2011) that may also, usefully, be sympathetic to the idea that cultural
objects are particulate in nature, as explained below.

Memes and cultural evolution

Names, in our construction, are posited as particles or units of culture, in
keeping with recent observations and empirical work in the application of
insights from evolutionary biology to human culture and social systems (Eades,
2015; Runciman, 2009; Dennett, 1995; Boyd and Richerson, 1985, page 37;
Dawkins, 1976). Within this cultural evolutionary framework, names as
memes (or particular units of culture) can be transmitted through both space
and time. Movement of names in space means the names are horizontally
transmitted (Eades, 2015) from person to person or from database to database
(and this is sometimes what is meant by the idea of 'going viral'). When names
are transmitted through time (i.e. handed down, remembered, performed orally
through inclusion in stories told to children or inscribed in old documents) they
are said to be vertically transmitted (Boyd and Richerson, 1985; Cavalli-Sforza
and Feldman, 1981). Though we do not use a quantitative approach in this
book, instead favoring qualitative examinations of geographical names, the
horizontal/vertical distinction is useful in adding nuance to the basic Kripkean
observation that names are communicated through chains of reference situated
in communities of practice (Kripke, 1981, pages 91–97; Evans, 1982, pages
373–404).

There are, furthermore, instances where we can be present (or nearly so) for
occasions of so-called 'pristine' naming, or the baptismal (originating)
moment of a name (Nash, 2013, page 6). In such cases, a feature in a landscape
goes from having no name to having one, without replacement of a prior
name. Unlike in colonial situations, where indigenous names are replaced by
the colonizer's names, in a 'pristine' case a feature in a landscape is quite

literally a blank slate. Nash's (2013) observations and ethnolinguistic considerations of such cases on Norfolk Island are instructive, and provide empirical and, importantly, geographical corroboration of Kripke's (1981) observation that names have baptismal moments (referred to by Wittgenstein (2009) as naming by ostension). Individual persons are, obviously, named at clearly identifiable points in time, but that this is also the case for geographical locales or objects is less apparent because such objects are often fuzzily defined (in space) or are named in ad hoc ways (Wittgenstein, 2009; Hough, 2015), indicating, in addition, a kind of temporal fuzziness or lack of clear temporal delineation.

Consider the area of British Columbia now referred to as the 'Great Bear Rainsforest' (GBR) (this example will be explored again below). As Wittgenstein has pointed out, one can point to a blob of matter and name it something (i.e. demonstratively, or by ostension, as it were). This is exactly what happened in the case of GBR. Environmentalists referred to a large area of temperate rainforest as The Big Green Blob before they knew what to call the part of coastal British Columbia they wished to save from clearcut logging practices (Page, 2014). Page, in his book *Tracking the Great Bear*, explores the process of coming to define, name, and map the so-called blob as 'GBR' through the use of actor-network theory (ANT), especially as theorized by Latour (2005). The ANT approach, though useful, is consciously eschewed in the present work precisely because it distributes agency too freely, resulting in overly diffuse networks that would seem to include almost anything. Furthermore, ANT-style networks often overlook what, for our purposes, is most interesting in the use of ANT, namely, naming and names (with Page's book actually representing a notable exception to this trend).

Through the use of ANT, Page (2014, pages 38–43) analyzes the origin and evolution of the name 'Great Bear Rainforest', exploring its ultimate referent and purpose by tracking its historical origins and contemporary politics. For our purposes GBR represents not only a case of pristine (and ostensive) naming, but a case of how maps are used as parts of name-tracking networks (NTNs) in chains of communication from an originating moment in time, with subsequent spread through space (i.e. in both vertical and horizontal transmission senses) (Hanna and Harrison, 2004). Here, communication and transmission are used almost synonymously and assume that there are senders and receivers of information, with that information kept alive and in its original form through four steps or salient criteria that ensure successful vertical transmission (through time) and horizontal transmission (through space). In this way, the name both 'sticks' (temporally) and becomes widespread (spatially). First, information (i.e. a name) must be successfully received (i.e. by the receiver of the information, whether a person, a document, or a memetic performance); second, the source and copy of the information (name) must resemble each other in relevant ways (the copy must be true to its source for the name to stick); third, that resemblance must be attributable to the source; and fourth, source and copy must co-exist or overlap for a period of time (the amount of time that will allow the copy to exist indefinitely) (Aunger, 2007, page 601 cited in Eades, 2015, pages 36–37). Thus, in four steps, transmission (and evolution) of names can be said to occur, regardless of the media (inscribed, oral, performed) used to facilitate it. Note that this cultural-evolutionary and memetic accounting for the mechanisms productive of a communication chain that preserves (adaptations of) names through time is in no way causal. The mechanism preserves the chain but none of its links are given impetus by cause and effect (Burgess, 2013, page 32)

For example, maps are parts of NTNs that facilitate successful vertical name transmission (i.e. the creation of communication links that propagate or propel the name through time). In the case of GBR the use of maps gives confidence (e.g. to environmentalists) that the name will 'take' or stick, which is another way of saying that it is here (or there) to stay (Page, 2014). GBR's appearance (always in the same place and covering the same extent) on official maps and in popular media also means that part of the effectiveness of the name is its ability to become a meme in the popular sense (i.e. it will stick in part because of the ease with which it can be horizontally transmitted through media other than, or in addition to, maps such as TV, radio, or newspapers). 'Great Bear Rainforest' as a thing is very new, but it remains a test case for how a name that claims a lot both discursively (politically), and spatially (covering a vast area) can come to pass (be baptized) in an ad hoc manner, but that subsequently spreads and captivates the imaginations of motivated populations and, thus, becomes a 'trackable' (real) part of a network.

Being in place

Names, despite their persistence and accuracy as (more than) labels for things, are also often in motion, moving from referent to referent over time and across space. While indigenous names are characterized by their persistence in place over time (Malpas, 2011), a name like 'pâté chinois' is derived from the town of China, Maine where the dish shepherd's pie was imported from England. Its original form was modified to include vegetables such as corn, readily available through farming by indigenous North Americans. With French and Quebecois adoption of the dish, the so-called 'China pie' became pâté chinois, a name any Quebecois will recognize as referring to a dish indigenous to their province (Blais, 2012). This leaves aside how a place in Maine came to have a name like 'China' in the first place and what this might mean (there is a connection to church hymns ...).

As the pâté chinois example points out, names are mixed beasts, and their geographies come in mixed forms and formats. This does not reduce—on the contrary, it only increases—the utility of NTNs in their tracking, as the use of an NTN to unpick the origin of the name pâté chinois points out. It can, however, present problems for the idea that indigenous names are somehow fixed or frozen in the past. Thornton (2012) has demonstrated that new indigenous names are being 'discovered' all the time and that, furthermore, many of these names are truly 'new' in the sense given above as pristine. We are present in the originating moment of some of these names that have arisen in response to some current or urgent need (Cruikshank, 2005), perhaps due to changing climate or colonization.

Plato (1926, page 19) states (speaking through the dialogue of Hermogenes and Socrates) that

SOCRATES: Now Naming is a part of speaking, for in naming I suppose people utter speech.

HERMOGENES: Certainly.

SOCRATES: Then is not naming also a kind of action, if speaking is a kind of action concerned with things?

HERMOGENES: Yes.

Plato anticipated later philosophy of language and the broader conception of names developed by Frege that is used in this book. Plato also anticipates Wittgenstein in the view that names are like tools for acting upon the world

SOCRATES: Then in naming also, if we are to be consistent with our previous conclusions, we cannot follow our own will, but the way and the instrument which the nature of things prescribes must be employed, must they not? And if we pursue this course we shall be successful in our naming, but otherwise we shall fail.

(Plato, 1926, page 21)

Plato is arriving at the view (again through a Socratic dialogue) that names are not mere labels or tags. They are, instead, tools for arriving at a true view of things and, in a moral sense, assessing the good and the bad (e.g. of a man's character). Names get at a true essence of things, therefore, in keeping with, for example, Plato's famous scene in the cave in *Republic*. The essentialist view, applied strictly to the name rather than the thing named, is one taken by Kripke when he states that there is an *a posteriori* necessity (Linsky, 2011; Fitch, 2004) to a name like Aristotle. Aristotle would still be Aristotle even if he had been born a different person (e.g. a musician). But he was not, so Kripke (1981) resorts to talk of possible worlds in which other Aristotles might have occurred (e.g. he might have been born brain-dead). Through possible worlds, Kripke avoids not only strict essentialism but also the circularity of names being defined in terms of other names (which occurs with a theory of language based on descriptions) (McGinn, 2015). It is the chain of communication of the name Aristotle from speaker to speaker that preserves its essence, even if Aristotle had been a very different person and even if other Aristotles were named over time. NTNs allow these different chains to be unpicked and mapped using a variety of representations, documents, and histories.

But if names are both particulate and mobile, where does this leave us vis-à-vis essentialism? To return to a previous example: what, essentially, is Greece; to what does it refer; how do we track the name using an NTN; how long has the name been there? What does it mean to be a Hellenic Republic or the birthplace of democracy? What do these things mean now, with, at the time of writing, Greece reconsidering its place in Europe and forging new alliances due to its perceived lower status among nations? There is a moral aspect to the name Greece, and it depends on both the time frame in which we choose to examine the name and the spatial frame with geopolitical implications. Due to a poisoned relationship dating from at least the Second World War, Greece is now at odds with Germany and has shown a willingness to align its interests more closely with Russia. A totalitarian strain is beginning to infect Greece's collective actions (or those of its current leader Tsipras) that at times seems to give the lie to its claim to be true to its democratic roots in Athens. The latter (a city, as opposed to a state) would probably have a stronger claim to the title (Brett, 2014) of birthplace of democracy. All this is to say that it is the name itself (e.g. Greece), as opposed to the thing named, which keeps it and associated knowledge alive (viable).

With all of this in mind and returning again to indigenous geographical names, Thornton (2008) has explored in detail how indigenous place names are different from non-indigenous place names. While Thornton's sophisticated theoretical framework for exploring Tlingit place names avoids essentialism through detailed treatment of how indigenous toponymies fit into greater kinship and social structures such as the potlatch, questions remain around the mobility and agency of the names themselves in geopolitical and cultural landscapes evolving through, and thus influenced by, American and Canadian state structures. As alluded to at the end of the introduction to this book, names are being used in creative new ways for indigenous efforts to reconnect to tradition and landscape. Such landscapes are not merely traditional, however. Through reforming processes such as the TRC in Canada, the idea of naming as anchoring of indigenous identity has both negative and positive consequences. The power to name a historic abuser is a negative identification or connotation in that the name is associated with trauma. Opposed to this is the positive idea of a (land-based) name as a way of overcoming trauma by (re)connecting with land on which ancestors derived livelihoods before contact with white outsiders (Eades, 2015). The latter, positive, connotation associated with the geographical name, whether new or old, represents an essential aspect of indigenous identification and healing.

Part of the approach of indigenous communities in Canada seeking land and place name-based ways of healing through reconnection is the location and (re)appropriation of colonial-era maps. Often these maps are used to overlay the old names over those dating from the colonization of an area (introduced by explorers and traders, for example) or over contemporary maps produced by the state. Place names introduced by early-contact explorers or merchants have found their way into contemporary and modern maps through chains of inscription and re-inscription in map production. In the example explored below, that of Boas (1969) and his famous 'Kwakiutl' geographical names, it is academia that can be implicated in a kind of soft colonialism—one that, I argue below, is ripe for re-appropriation by Kwakwaka'wakw (the current transliteration of the word formerly referring to 'Kwakiutl') peoples of what is now central coastal British Columbia, themselves (Galois, 1994).

'Kwakiutl' geographical names

It is instructive to revisit Boas's (1969) classic of geographical name studies, *Geographical Names of the Kwakiutl Indians*. In this work, Boas covers aspects of indigenous place-naming not only among the 'Kwakiutl' (Kwakwaka'wakw) but also amongst the Tewa, 'Eskimo' (Inuit), and Navajo, to name just three other indigenous groups whose cultural and linguistic attributes are examined in comparison to the Kwakiutl 'Indians' (as they were known at the time). Boas's methodology distinguishes between cultural and linguistic practices of place-naming. The former, for the Kwakiutl, indicate *what* is named, with several distinct categories in evidence. Thus, features in land and seascapes of cultural salience and hence likely to acquire names are those important for navigational purposes, with an orientation relative to coasts and rivers. A focus on transportation and waterways is a recurrent feature not only in northern Canadian indigenous life-worlds and naming systems (Eades, 2012) but also in North American indigenous systems more generally (Meadows, 2008).

Other cultural aspects of Kwakiutl geographical names begin to hint at structures and attributes of indigenous names more generally (Afable and Beeler, 1997). Part of the cultural aspect of geographical names is their use value as tools for locating valuable resources such as fish or berries. Names can in this way indicate areas or sites acting as receptacles or containers of food including sockeye salmon, trout, clams, fish eggs rich in nutrients, blueberries, elderberries, and high-value cedar bark. Settlement names indicate where food is likely to be processed, often referring to the foundations on which the village sits. Rock, beach, and log foundations are thus referred to in Kwakiutl naming systems, giving a literal aspect of truth to the idea that indigenous names are foundational in nature. What is meant by this—and there are implications for the trajectory of this book—is that, philosophically speaking, traditional naming systems had an interest in referring straightforwardly, and often quite descriptively, to materials or values contained in the referent. The materiality of names comes through in Kwakiutl geographical names that indicate the kind of rock forming the foundation of the place or are obvious in the appearance of the place (Boas, 1969).

Language, on the other hand, is demonstrated by Boas to constrain what can be named or the style in which this occurs. He does so by comparing the structures of Kwakiutl and 'Eskimo' languages. The former contains a great number of suffixes that can be attached to other words or parts of words to form new names. The latter relies more upon whole stem words that qualify or are qualified by other words. We have noted the use or tool value opened up by new inquiry into Boas's *Geographical Names* text and his examination of the Kwakiutl culture and language in relation to others. What possible worlds does such an inquiry open up? Language as a determinism would almost seem to point to chorological or regional facts differentiated by area in Sauer's (1963) sense. Kripkean, post-foundational geographies of possible worlds would not be held to such determinisms about the shape of things in the world, nor about the discipline of geography itself. What the Kwakiutl geographical names of Boas indicate is that a group of individuals was sectioned off as distinct through the use of categories of names for descriptive, locative, navigational, proximal, cardinal, dwelling, resource, and other purposes. A colonial mindset is not immediately apparent until one arrives at the maps, which simultaneously act as colonizing tool and NTN, at the back of Boas's book. Here we see a grid laid out bare with the facts. A potential Kwakiutl gazetteer presents itself for mapping and for fixing Kwakiutl in place, separate from other sets of phenomena that lie beyond the edges of those maps. The colonizing aspect of Boas's (and others of his time and disciplinary bearing) anthropology is often overlooked in studies that accept lists and grids as objective facts, without noting how that same objectification freezes out consequent evolutions, trajectories, or additions to the map/gazetteer. The assumption is one of loss, erosion, or, at best, residual relevance in the archive.

To put this into some kind of context, we need to consider what bearing the observations of Boas or Sauer had on residential schooling systems in Canada that lasted throughout most of the twentieth century. The essential background is that knowledge was colonized, fixed as fact, placed in museums, institutionalized behind the backs of those to whom original meanings were the phenomenal living heart of the matter. That things could have been otherwise is as historical and contingent a fact as to say that Boas's maps might have looked much different had he arrived in a different century. His fixing upon the page of the geographical names of a whole people was part and parcel of his (and the scientific establishment's) worldview. It clashed with those under study, for whom the knowledge on display was fundamentally oral and performed in nature, and thus both transmitted and open to change in a fundamentally different and less catastrophic way. But that fact was swept under the rug for over a century. Kwakiutl is therefore a name with colonial overtones, a negative fact in itself, an *a posteriori* necessity for turn-of-the-century anthropology, but one that now looks a bit antiquated and certainly does not do justice to the fullness of Kwakwaka'wakw knowledge, worldviews, and place-worlds (Galois, 1994).

Current anthropologies of northern Canada accept the fact of colonization, incorporating the emic (insider's take) into their purview. Niezen (2013, pages 112– 113) has demonstrated that another world is possible for non-indigenous academic minds through the use of sophisticated new ways of ethnographic inquiry, noting in relation to the TRC that

> The main difficulty here is that the Commission's guiding principles and assumptions are already so familiar that it is nearly impossible to

disentangle them from widely current ideas (or unknowable hegemonies, if one wanted to push the difficulty to its limit)—from the ideological air we breathe. This is where the pioneering ethnographers of, say, British social anthropology had it much easier. When Malinowski wanted to understand the kula exchange of the Trobiand Islanders, or Evans-Pritchard the logic of witchcraft among the Azande, their task, as they saw it, was to translate deeply unfamiliar ideas into the vernacular of their readership, to make the seemingly mysterious interpretable and the incommensurable commensurable. The effort to explore the social workings of ideas that are already at some level deeply familiar, like those at the foundation of the TRC, is an altogether different challenge.

We see in Niezen's observation how much is at stake in any analysis of names with the potential to reconnect traumatized populations with their lands, and how far ethnography and anthropology have come to define what is possible in indigenous life-worlds in Canada and beyond.

Nunavik

In the winter of 2012 I was part of a team sent to northern Quebec to interview Nunavik elders. Nunavik refers to the traditional Inuit portion of Quebec. In the south its boundary includes the northern part of James Bay where it edges against Cree territories, starting around Kuujjuarapik (Great Whale). Both Kuujjuarapik and Chisasibi (a primarily Cree town) have Inuit and Cree populations and elders will tell stories about wars fought in southern Nunavik, at times when food was scarce. The southern boundary of Nunavik extends across the Ungava peninsula towards the modern boundary between Labrador and Quebec and then northwards to the Hudson Strait. Nunavik includes Ungava Bay and a large part of the Hudson Bay coastline, but very little of Hudson Bay itself, which was assigned to the Inuit of the newly formed territory of Nunavut to the north. There are fourteen Inuit communities in Nunavik, from Killiniq in the extreme east to—following the coastline, where all Nunavik communities are situated—Kuujjarapik, farthest to the west, on Hudson Bay.

  Armed with hundreds of maps, my task was to verify place names written in a transliterated form of Inuktitut for the Nunavik Inuit Land Claims Agreement (NILCA). NILCA has provisions for islands belonging to the territories of Nunavik Inuit, in places where the land–sea interface is complex and shows a clear and historical association with the hunting, fishing, and gathering activities of Quebec Inuit as opposed to Inuit from more northern locales belonging to Nunavut. I was thus part of an internal sorting process involving two separate Inuit groups and the governments of Canada and Quebec, both of which have an interest in naming the islands in either English (Canada) or French (Quebec), not to mention the native languages. I visited eleven of the fourteen communities, starting in Kangiqsualujjuaq and moving west along the shoreline through the 'capital' of Nunavik, Kuujjuak, and onward to

Tasiujaq during the first leg of the journey in February. The structure of our meetings with elders was such that a call was put out before our arrival announcing the time, place, and purpose of our meeting. Meetings usually included between one and ten elders, a mixture of men and women. The only condition for an elder's inclusion as participant or informant at a meeting was extensive knowledge of their home territories, and this was usually obtained through a lifetime of hunting and trapping on the land and sea.

The meetings were informal and always began with introductions and usually a few observations on the weather as coffee was served. Food and drink were very important parts of the meetings, showing good will and blunting the formal edge of the occasion. We knew the meeting had started when the maps were pulled out. The maps had been produced through the use of a database developed by Muller-Wille of McGill University's department of geography, along with the cartographic production facilities of a consulting firm, also in Montreal, Quebec. A key map helped us situate ourselves, usually starting (on the maps) somewhere near the settlement, moving outwards alongshore or upstream as appropriate. Discussions would often veer toward the subject of proposed mining activity. Other times we would talk about the large islands situated offshore, such as Akpatok, Mansel, and Nottingham. These islands tended to be associated with colonial-era stories of explorers, missionaries, or evil spirits, to give just three examples. A famous murder had occurred on Mansel Island, for example, when an unfortunate Inuit man had been possessed by an evil spirit. The stories, in this instance, were told by Ivujivik elders and took up a good portion of one of our four days' worth of meetings in that community.

It's important to point out that the testimonials, narratives, and place names verified, added, or deleted from the maps were legally binding, signed by participants and witnessed by me, a translator, and an assistant, the latter two being themselves representatives from the association of Quebec Inuit elders, the Avataq Cultural Institute. This kind of hearing was legitimated in the eyes of Canada by my status as a credentialed academic, and by the community-sanctioned knowledge of the elders. Elders in Nunavik Quebec have formed the above-mentioned ('Avataq') association to protect their knowledge, with offices for convenience in Montreal. The names of respected elders in Inuit Quebec carry as heavy a weight as the name of a judge might carry in a courtroom in more southern parts of the country. The legitimating of elder involvement in negotiating the boundaries between outsiders such as myself and the on the inside of Inuit communities is of exceptional importance, and only achieved on these visits through the help of my assistant-translator, a young Inuit man fluent in three languages. This young man contacted elders in all eleven communities, and the fact that he spoke Inuktitut fluently meant he could open doors that I, with my lack of knowledge of the language, could not.

The kind of truth that was conveyed through the use of maps in conversation with Nunavik elders is on the face of it quite different from that conveyed through something like a TRC meeting. But below the surface differences, one finds that they are in fact two sides of the same coin. As Niezen (2013, page 132) notes, there are "connections between residential schools and territorial removal as well as the traumas associated with the school experience." Many elders brought up their school experience, noting good times and bad, the bad mostly trauma associated with being away from home for the first time and with being forbidden, as was government policy, from speaking their own language. One elder talked about being shipped away for tuberculosis treatment. Most of the time, however, 'mere' talk of place names, the land, and associated stories was seen as a healing and reconnecting process.

Several issues arise when considering Nunavik place names and their continued viability. While land use and occupancy studies and place-name verification surveys continue unabated to the present day, often led by non-local and/or non-indigenous individuals, their relevance to populations being surveyed is threefold (Tobias, 2009): it is to do with *continuity, identity*, and *representation*. Kishigami (2006, page 206) has argued persuasively for several factors contributing to *continuity* for Inuit living in urban settings, noting that "Inuit kinship … is not merely a matter of biological relatedness but depends on many other factors, such as physical proximity, naming, and regular social interactions." Collignon (2006) has similarly argued for continuity through place-naming and its relationship with traditional and land-based ways of life. Thus there are two sides to the continuity coin, one corresponding to centralization and urbanization, with attendant informal and formal social networks, and the other to efforts to keep alive or re-enliven Inuit interactions with the land and sea of traditional territories. By keeping alive frameworks and knowledge structures such as those contained in gazetteers produced by land use and occupancy studies and place-name surveys, outsiders are working in partnership with indigenous and Inuit peoples to promote continuity of culture (Muller-Wille and Muller-Wille, 2006; Muller-Wille, 1987; Mead, 1964).

In terms of continuity, place names and knowledge attached to those names provide tools for producing possible worlds that are viable for the greatest number of Inuit. Temporally, living knowledge produces more subjects as it accumulates knowledgeable individuals interacting with traditional and evolving knowledge systems. This does not preclude the coining of new place names that represent the state of the art in knowledge but also tradition and practice (that of geographical naming). New names are constantly being coined as need arises. Others, less used, may fall into disrepair, become difficult to translate due to their association with old ways of speaking, or disappear altogether. In the main, however, landscapes opened up by names in the north (of Quebec as we see here) have a consistency, coherence, and a community of practice that keeps them alive and constantly hone their usefulness as tools of language. Place names are literally posts at which we station words (Wittgenstein, 2009, page 18) and this is especially both literal and true in Inuit and indigenous life-worlds in Canada.

Against colonial place-naming in Canada, which often came up in conversations with elders in Quebec, Inuit *identities* counter-map dominant or hegemonic (Canadian) identities. For example, many offshore islands in Nunavik have English names. In conversation with elders, the perception is that these are names introduced by Ottawa, meaning the Canadian government. But the government is clearly not the ultimate origin of the names introduced by early explorers of the region in many cases (e.g. Mansell or Digges islands near the community of Ivujivik). Other islands that somehow retain their Inuit names, such as Akpatok Island, are perceived to have been colonized by their inclusion on an English map that transliterates Inuktitut. The latter (Inuktitut) is included for ease of use by outsiders such as myself but such a map, containing place names spelled out literally (i.e. in English as opposed to using the syllabic script preferred by Inuit and Cree) is considered by many elders to be an outsider's map, despite containing Inuit place names. Thus, there are two levels to spatial identities, that of language, and that of content. This aspect of identity is explicitly spatial in the sense that it is tied to paper maps (preferred in contexts where work is with elders who are usually considered 'old school' and thus prefer large, hands-on, full-spread paper maps as opposed to screens), with their top-down, god's eye (and thus attractively authoritative) view of things.

This aspect of spatial identity is contrary to many geographers' and anthropologists' sense of indigenous knowledge systems, which are often constructed as being strictly grounded, phenomenological, and embodied. Through time, however, and in interaction with 'outside' representatives from trading companies, hunters, and government officials (as well as academics), local Cree, Inuit, and indigenous communities in Canada have adopted maps as a preferred format for discussing, validating, and legitimizing observations and named spaces. As noted above, these maps are signed by elders in the presence of expert witnesses and then deposited in government offices for digitization, inclusion in databases and on maps, and ultimately archiving. The tools in play here are those associated with utterance and truth, different ways of truth-telling (oral testimony), and styles of discourse. The way one tells a spatial story over a map is different from how one would tell the story without a map. The top-down, god's eye view explicitly links places to each other in a way that simple utterance does not.

Thus, place-based identity forms another aspect of indigenous knowledge systems, and these are tied to the grounded view that complements that noted above (the god's eye view). Place identity has been well covered in the geographical literature (Cresswell, 2012) and will not be covered in depth in this chapter, but it is an important thread that runs through this book. In terms of maps and geographical names, there are specific pathways for tracking names through time, space, and the human body, and they are intimately caught up with issues of *representation* (Malafouris, 2013; Taylor, 2012; Ismael, 2007). Naming performances enacted through both utterance and interaction with named places on the land are toolboxes that also include ritual aspects. Place-naming follows well-established pathways originating simultaneously in brain-based spatial structures, bodily ways of 'being there', discursivity, and practices. Ultimately, each individual has a representing mind, and bodies that interact with external spatial inscriptions such as maps re-inscribe internal representations each time an interaction takes place either between an *agent* in a landscape or with several agents telling stories over maps, through stories, or even just in remembering things past.

These will be controversial claims and they need much unpacking. I devote the next several chapters to doing so.

References

Aunger, Robert. 2007. Memes, in Dunbar, Robin and Barrett, L. (eds). *Oxford Handbook of Evolutionary Psychology*. Oxford: Oxford University Press.

Afable, Patricia O. and Beeler, Madison S. 1997. Place-Names, in Goddard, I. (ed.). *Handbook of North American Indians*. Vol. 17, *Languages*. 185–189. Washington: Smithsonian Institution.

Blais, Christina. 2012. Pâté Chinois 101. *Ricardo*. 10(1). 17–19.

Boas, Franz. 1969. *Geographical Names of the Kwakiutl Indians*. Columbia University Contributions to Anthropology, vol. 20. New York: Columbia University Press.

Boyd, Robert and Richerson, Peter J. 1985. *Culture and the Evolutionary Process*. Chicago: University of Chicago Press.

Brett, Annabel S. 2014. *Changes of State: Nature and the Limits of the City in Early Modern Natural Law*. Princeton: Princeton University Press.

Burgess, John P. 2013. *Kripke*. Cambridge and Malden: Polity.

Cavalli-Sforza, Luigi and Feldman, Marcus. 1981. *Cultural Transmission and Evolution: A Quantitative Approach*. Princeton: Princeton University Press.

Clark, Andy, 2011. *Supersizing the Mind: Embodiment, Action, and Cognitive Extension*. Oxford: Oxford University Press.

Collignon, Beatrice. 2006. *Knowing Places: Innuinnait, Landscapes and Environment*. Calgary: CCI Press.

Cresswell, Tim. 2012. *Geographical Thought: A Critical Introduction*. Chichester: Wiley-Blackwell.

Cruikshank, Julie. 2005. *Do Glaciers Listen? Local Knowledge, Colonial Encouters, and Social Imagination*. Vancouver: UBC Press.

Dawkins, Richard. 1976. *The Selfish Gene*. Oxford: Oxford University Press.

Dennett, Daniel. 1995. *Darwin's Dangerous Idea*. New York: Touchstone.

Eades, Gwilym. 2012. Cree Ethnogeography. *Human Geography*. 5(3). 15–31.

Eades, Gwilym. 2014. Toponymic Constraints in Wemindji. *The Canadian Geographer*. 58(2). 233–243.

Eades, Gwilym. 2015. *Maps and Memes: Redrawing Culture, Place, and Identity in Indigenous Communities*. Montreal and Kingston: McGill-Queen's University Press.

Evans, Gareth. 1982. *The Varieties of Reference*. New York and Oxford: Oxford University Press.

Fitch, G.W. 2004. *Saul Kripke*. Chesham: Acumen.

Galois, Robert. 1994. *Kwakwaka'wakw Settlements: 1775–1920*. Vancouver: UBC Press.

Gelling, Margaret and Cole, Ann. 2000. *The Landscape of Place-Names*. Stamford: Shaun Tyas.

Glacken, Clarence J. 1967. *Traces on the Rhodian Shore: Nature and Culture in Western Thought From Ancient Times to the End of the Eighteenth Century.* Berkeley and Los Angeles: University of California Press.

Hanna, Patricia and Harrison, Bernard. 2004. *Word and World: Practice and the Foundations of Language.* Cambridge: Cambridge University Press.

Hawkins, Harriet and Straughan, Elisabeth (eds). 2015. *Geographical Aesthetics: Imagining Space, Staging Encounters.* Farnham: Ashgate.

Hough, Carole. 2015. Places and Other Names, in Taylor, John R. (ed.). *The Oxford Handbook of the Word.* Oxford: Oxford University Press. 634–649.

Hunn, Eugene. 1994. Place-Names, Population Density, and the Magic Number 500. *Current Anthropology.* 35(1). 81–85.

Hunn, Eugene. 1996. Columbia Plateau Indian Place-Names: What Can They Teach Us? *Journal of Linguistic Anthropology.* 6(1). 3–26.

Hunn, Eugene and Meilleur, Brien. 2010. Toward a Theory of Landscape Ethnoecological Classification, in Johnson, Leslie Main and Hunn, Eugene (eds). *Landscape Ethnoecology: Concepts of Biotic and Physical Space.* New York and Oxford: Berghahn.

Ismael, J.T. 2007. *The Situated Self.* Oxford: Oxford University Press.

Johnson, Leslie Main and Hunn, Eugene (eds). 2010. *Landscape Ethnoecology: Concepts of Biotic and Physical Space.* New York and Oxford: Berghahn.

Kishigami, Nobuhiro. 2006. Inuit Social Networks in Urban Settings, in Stern, Pamela and Stevenson, Lisa (eds). *Critical Inuit Studies: An Anthology of Contemporary Arctic Anthropology.* Lincoln: University of Nebraska Press.

Kripke, Saul. 1981. *Naming and Necessity.* Malden: Blackwell.

Latour, Bruno. 2005. *Reassembling the Social: An Introduction to Actor-Network Theory.* New York: Oxford University Press.

Levi-Strauss, Claude. 1966. *The Savage Mind.* Chicago: University of Chicago Press.

Levi-Strauss, Claude. 1983. *The Raw and the Cooked.* Chicago: University of Chicago Press.

Linsky, Bernard. 2011. Kripke on Proper and General Names, in Berger, Alan (ed.). *Saul Kripke.* Cambridge: Cambridge University Press.

Malafouris, Lambros. 2013. *How Things Shape the Mind: A Theory of Material Engagement.* Cambridge, MA: MIT Press.

Malpas, Jeff (ed.). 2011. *The Place of Landscape: Concepts, Contexts, Studies.* Cambridge, MA: MIT Press.

Mark, David, Turk, Andrew, Burenhult, Niclas and Stea, David (eds). 2011. *Landscape in Language: Transdisciplinary Perspectives.* Amsterdam: John Benjamins.

McGinn, Colin. 2015. *Philosophy of Language: The Classics Explained.* Cambridge, MA: MIT Press.

Mead, Margaret. 1964. *Continuities in Cultural Evolution.* New Haven: Yale University Press.

Meadows, William. 2008. *Kiowa Ethnogeography.* Austin, TX: University of Texas Press.

Muller-Wille, Ludger and Muller-Wille, Linna Weber. 2006. Inuit Geographical Knowledge One Hundred Years Apart, in Stern, Pamela and Stevenson, Lisa (eds). *Critical Inuit Studies: An Anthology of Contemporary Arctic Ethnography.* Lincoln: University of Nebraska Press.

Muller-Wille, Ludger. 1987. *Gazetteer of Inuit Place Names in Nunavik, Quebec, Canada.* Montreal: Avataq Cultural Institute.

Nash, Joshua. 2013. *Insular Toponymies: Place-Naming on Norfolk Island, South Pacific and Dudley Peninsula, Kangaroo Island.* Amsterdam: John Benjamins.

Niezen, Ronald. 2013. *Truth & Indignation: Canada's Truth and Reconciliation Commission on Indian Residential Schools*. Toronto: University of Toronto Press.

O'Keefe, John and Nadel, Lynn. 1978. *The Hippocampus as a Cognitive Map*. Oxford: Clarendon Press.

Page, Justin. 2014. *Tracking the Great Bear: How Environmentalists Recreated British Columbia's Coastal Rainforest*. Vancouver: UBC Press.

Plato. 1926. *The Works of Plato*. Vol. VI, *Cratylus*. London: William Heinemann.

Rowlands, Mark. 2010. *The New Science of Mind: From Extended Mind to Embodied Phenomenology*. Cambridge, MA: MIT Press.

Runciman, Walter G. 2009. *The Theory of Cultural and Social Selection*. Cambridge: Cambridge University Press.

Sapir, Edward. 1958. Time Perspective in Aboriginal American Culture: A Study in Method, in Mandelbaum, David (ed.). *Selected Writings of Edward Sapir*. Berkeley and Los Angeles: University of California Press.

Sauer, Carl. 1963. The Morphology of Landscape, in Leighly, John (ed.). *Land and Life: A Selection From the Writings of Carl Ortwin Sauer*. Berkeley and Los Angeles: University of California Press.

Taylor, Kathleen. 2012. *The Brain Supremacy: Notes from the Frontiers of Neuroscience*. Oxford: Oxford University Press.

Thornton, Thomas F. 2008. *Being and Place Among the Tlingit*. Seattle: University of Washington Press.

Thornton, Thomas F. (ed.). 2012. *Haa Leelk'w Has Aani Saax'u/Our Grandparents Names on the Land*. Seattle, London and Juneau: University of Washington Press/ Sealaska Institute.

Tobias, Terry. 2009. *Living Proof: The Essential Data Collection Guide For Indigenous Use-and-Occupancy Map Surveys*. Vancouver: Ecotrust Canada.

Turner, Nancy. 2014. *Ancient Pathways, Ancestral Knowledge*. Montreal and Kingston: McGill-Queen's University Press.

Wittgenstein, Ludwig. 2009. *Philosophical Investigations*, 4th edition. Chichester: Wiley-Blackwell.

3 Religion and geographical naming

This chapter outlines several examples of associations between geographical naming and religion within a wide-ranging spatial and temporal framework. I cast my net very wide indeed across time and space and various scales in order to make the point clear that aspects of naming associated with belief, magic, religion, and ritual can all be accommodated within communication- and tool-based ways of thinking. The scholarly materials consulted below form parts of NTNs for both fixing and propagating essential qualities of place names for future use and study and, as such, are treated as primary materials despite the majority containing commentaries and editorials that are more logically interpreted as secondary. For example, both the Mayan (Stuart, 1994) and Welsh (Jones, 1954) points of reference contain lists of toponyms associated with religious beliefs and practices. The Humboldt (2012) material is especially concerned with such practices as they revolve around particular named places in South America, namely, volcanoes. Material concerned with beating the bounds and rogation in their modern-day forms was collected from the Bodleian library and represents folk inscriptions of place-based and spatial religious practices.

In addition to the primary framing texts of Kripke (1981), Wittgenstein (2009), and Hanna and Harrison (2004), this chapter makes limited use of Mauss's (2001) work on magic. Connections between magic and names will become clearer below but some comments here serve to introduce the new conceptual scaffolding. Name-bearers hold contradictory beliefs about names, which, like many phenomena, are closely connected to basic consciousness and ways of being, both objective and subjective (Nagel, 1986). We tend to believe that names both reflect what's in the tin and, somewhat contradictorily, act as mere tags or labels. The two ways of seeing cannot be always true at the same time—or it can be said there is overlap between the two claims that we might call *correspondence* (name reflects what's in the tin) and *arbitrariness* (label is just an attached tag with no bearing on contents). If proper names truly reflect what's in the tin (the more objective claim), one must ask: how is this achieved? It is an especially salient question when applied to person names (which admittedly are not usually very descriptive), but it is equally interesting in the case of geographical names.

Except in cases where there is a clear lineage, such as when a son is named after his father or, in a colonial context, when a place (e.g. Athens, Georgia) is clearly part of a meme communicating another idea (in this example, of the Athens of ancient Greece) through its name, the idea that a name reflects what's in the tin would seem both counterintuitive and magical. Even in the cases just mentioned, the fact that the name is a meme (i.e. a cultural object transmitted through time or across space) reflects something about the intentions of the one doing the naming rather than the contents of the thing named. If saying a name, then, is a bit like pulling something out of a hat, the reason for this is that names acquire *a posteriori* necessity through long usage and chains of communication over time. In fact, this can happen in less than a generation as the many Johns, Davids, and Sarahs will come to be associated uniquely (and subjectively) with their names, despite a plethora of others with the same name. The trick with names, and what gives them magical properties, is the power associated with their performance (utterance, inscription, or evocation). We ask here, in the case of names, is the one bestowing a name like a magician, installing future belief in the correctness of the name *a posteriori*, and what is represented through the god-trick of bestowing a name? Belief in names and their performance are ritualized events that benefit from repetition to create belief in their correctness after the fact. The utterance of a name is the ultimate attention-diverting activity in the name of representing the authority of the one both bearing and using the name.

For geographical names, as described below, magic and religion are inseparable, and belief is treated as an ethnographic fact (Blaut, 1979), worthy of study in its own right, on its own terms.

Mayan inscription of place

Mayan place-name glyphs combine image and text in hieroglyphic geographical references, and this makes them quite unique in the world of geographical names and naming systems where, for the most part, we tend to adopt a binary divide separating oral utterance from written inscription. The Mayan inscriptions are also visual, nudging them closer to the oral, sound-based, 'side' of the divide (i.e. the audio-visual). If a picture is worth a thousand words, a place glyph is worth at least as many. Indeed, the hieroglyphic structure of Mayan place inscriptions makes them not only unique but also flexible and compact, containing a good deal of information within a few small (and often variably orientated) marks. Description and analysis of Mayan place glyphs are included here, under this chapter's ostensible heading of religion, despite having demonstrated political value for assessing ancient indigenous power structures as outlined by Stuart (1994, page 3): "we believe that Emblem Glyphs refer to political units that could incorporate several named sites." Each Mayan place glyph, then, holds the potential to refer to several places at once, those falling under the power of a particular lord, deity, or powerful ruler.

It is worth reviewing Stuart (1994) in greater depth as a primary source and set of categories of geographical names useful in comparison with others covered below, and as a potential foundation for analysis of inscribed place names (i.e. at the opposite end of the spectrum from the audio-visual). Stuart (1994, page 17) notes that

> all writing systems are to one degree or another incomplete records of speech. Specialists in Sumerian cuniform (a logo-syllabic writing system that is typologically similar to that of Maya), commonly acknowledge the scribal underrepresentation of sounds that "should" be there.

This refers to the fact that Mayan place glyphs often use *u-ti-ya* as a prefix to the specific named place, in what Stuart calls the place name "formula" (Stuart, 1994, page 8). *U-ti-ya* translates roughly as 'it happened,' in essence dropping the 'at' in the process. Stuart's publication is in large part a defense of the interpretation of 'emblem' and other glyphs as referring specifically to places despite this missing word ('at'). In providing evidence or arguments supportive of a place-glyph thesis and in addition to a working list of Maya place glyphs, Stuart (1994) produces the following categories: "Place Names with Other Verbs," "The Iconography of Place Names," "Mythological Place Names," followed by site area descriptions and two case studies.

It is noteworthy that Stuart refers to the place names first and the glyphs second, in keeping with a general emphasis on material culture. External referents (i.e. in the world) are demonstrated to have consistent references (names) according to the place-glyph formula, often with religious, mythological, or political significance. The so-called emblem glyphs that include the name of a lord or other powerful figures are explained as being distinct and separate from place glyphs following the *u-ti-ya* formula. Many of the glyphs refer to water bodies ending with the word particle 'ha' (for water), and this usage is noted as reflecting modern Mayan place names that often end in 'ha' (e.g. Yaxha). This continuity with modern practice is notable for two reasons. First, pronunciation of the water syllable demonstrates the lasting durability of toponyms from almost baptismal or pristine moments (prehistoric) to the present. Language as a visual (i.e. hieroglyphic) medium has evolved into the written script that we know today without diminishing that factor of continuity. Second is the issue of uncertainty of the boundary of the entity referred to by the geographical name. Contemporarily we can think of examples of where the referent of a name is, for example, a water body but the boundary of what is referred to (e.g. lake or river) remains uncertain. In the case of Mayan glyphs the following was observed: "even the *Yaxha'* decipherment of the Yaxha Emblem Glyph still fails to elucidate the geographical scope of its referent" (Stuart, 1994, page 7).

The same would be true, though often unchallenged, in a modern GIS, which would most likely delineate the boundary of the entity as corresponding strictly to the edge of the lake, with a hard (inscribed or fixed) edge.

Furthermore, this lake could be represented as an unchanging entity with fairly fixed levels. The name itself does something similar by fixing its referent, but it does so vaguely and even subjectively, depending upon the person uttering the name and his or her idea of what it corresponds to. However, precise delineation (unlike, perhaps, in a GIS) may be beside the point, as the subjectivity argument points out. Place names used in association with verbs including, but not limited to, *u-ti* show that it is an action element to the sentence as a whole that is of much more importance and has resulted in a scribe writing down the sentence in the first place. (Even the word 'at', for example, is usually seen as beside the point in these glyphs). Another important function of place names in Mayan life-worlds was to accompany personal names (Stuart, 1994, page 33). In these cases, the place names functioned as tools for anchoring both the identity and area of power with which a named (powerful) individual was associated. These title-of-origin names are again preceded by the language particle 'ah', in a kind of formula that is repeated over and over in the material record. For example, *Ah Wak'ab* refers to a lord (also referred to in an emblem glyph) associated with what in modern-day parlance is called the Usumacinta River (Stuart, 1994, page 36); another refers to *Lakamtun*, of unknown location, but the association of this glyph with a lord's emblem and with other place glyphs is given as evidence of reference to location (Stuart, 1994, page 37).

The supernatural (and thus religious) aspect of Mayan place names is noted by Stuart: "place names also appear in texts that treat mythological or supernatural happenings. These can be seen on many examples of Maya art from the Classic period and allow for several new insights into the geography of Maya religious belief" (Stuart, 1994, page 69). Mythological toponyms refer here to places associated with the doings of supernatural beings. *Na-Ho-Kan* means roughly five skies. "*Nahokaan Ahaw* ('lord[s] of Nahokaan') ... serves notice that the Paddlers [the names of these particular deities] were lords of a particular place in Maya mythic geography. Without the ahaw [lord] glyph, it functions simply as a place name," albeit one that no mere mortal would be able to find within his or her lifetime (Stuart, 1994, page 71). Mythical naming of this type takes toponymy into a whole new register. By fixing in (real or mythical) space the figures of powerful individuals, including lords and gods, their power is given additional legitimacy, beyond what the image of the powerful figure or the place name, taken separately, could produce. Looking ahead a couple of chapters, we can see that such toponyms are thus tools of power and politics. They act like metaphorical screwdrivers for attaching, anchoring, or affixing identity in space. In a case such as that of the lords above, where the referent is vaguely attached to skies, the power of the name is only increased, rather than the opposite (decrease in power due to diffusion over an area of the referent). The mythical site of the five skies takes us beyond the one sky we see with our eyes to others beyond mortal comprehension. Hierarchies are thus established and through the *ahaw* glyphs we see toponymic power consolidation at one remove from that of mythical lords in the sky. The power of the lords is similarly all-pervasive, but here we at least have place names everyone (in the Mayan world) knows and that are broad in spatial extent. Thus there is assertion of power over space and an early form of geopolitical wrangling apparent in the Mayan place-glyphs. Again we anticipate both cognitive and political aspects of names, here in the religious and mythical Mayan glyphs.

Naming, baptism, and ostension

Homer (2008) alludes several times in *The Iliad* to examples of people named after geographical entities originally named by gods. Socrates, in Plato's (1939) *Cratylus,* repeats many of these as examples of 'correct' names. They are considered correct insofar as they originate in impulses of the gods who have ostensibly carried out acts of naming based upon some underlying essence or truth of nature reflected in each name. For example, in Homer (2008, page 352)

> Leto was opposed by gracious Hermes
> wayfinder for souls, and the god of fire
> Hephaestus, facing a mighty eddying river,
> Xanthus to the gods, to men Scamander.
> These were the divine adversaries.

Plato (1939, page 35) points out that the true names are always there beneath the flux of war and the deeds of men, and that indeed men (though not, for Homer, women) are, after the gods, best equipped to know the true state of things and thus to name (page 37). On page 40, again, we have Homer (2008) proclaiming on the names of men, in relation to plains of war

> Rising in isolation on the plain
> in the face of Troy, there is a ridge, a bluff
> open on all sides: Briar Hill they call it.
> Men do, that is; the immortals know the place
> to be the Amazon Myrine's tomb.
> Anchored on this the Trojans and allies
> formed for battle.

To know the true name of an object is to be in a privileged position: to be a man or a god. Mortal philosophers position themselves on a continuum somewhere between the two. Thus, we have Socrates' oracular proclamations on names as parts (or particles) of language, with letters of the alphabet themselves acting as names that can be added together to refer to things in the world directly (Plato, 1939, page 41). Names can be read off from the world unproblematically, in keeping with one's privileged position vis-à-vis knowledge of metaphysics. Later, Wittgenstein (2001, page 15) proclaims knowledge of names in much the same vein: "The simple signs employed in propositions are called names … . A name means an object. The object is its meaning."

Hubris is attached to acts of naming, which seems to exhibit a pride beyond what has been attained. In Homer, men can do nothing against the river Scamander (Astyanax). The river in turn can do nothing in the power of the god Hephaestus who

> brought heaven's flame to bear: upon the plain
> it broke out first, consuming many dead men
> there from the number whom Achilles killed,
> while all the plain was burned off and the shining
> water stopped. As north wind in late summer
> quickly dries an orchard freshly watered,
> to the pleasure of the gardener, just so
> the whole reach of the plain grew dry, as fire
> burned the corpses. Then against the river
> Hephaestus turned his bright flame, and the elms
> and tamarisks and willows burned away,
> with all the clover, galingale, and rushes
> plentiful along the winding streams.
> Then eels and fish, in backwaters, in currents,
> wriggled here and there at the scalding breath
> of torrid blasts from the great smith, Hephaestus,
> and dried away by them, the river cried:
> 'Hephaestus, not one god can vie with you!
> Neither would I contend with one so fiery.
> Break off the quarrel: let the Prince Achilles
> drive the Trojans from their town. Am I
> a party to that strife? Am I their saviour?'
> He spoke in steam, and his clear current seethed
> the way a cauldron whipped by a white-hot fire
> boils with a well-fed hog's abundant fat
> that spatters all the rim, as dry split wood
> turns ash beneath it. So his currents, fanned
> by fire, seethed, and the river would not flow
> but came to a halt, tormented by the gale
> of fire from the heavenly smith, Hephaestus.
>   (Homer, 2008, pages 373–374)

The name here speaks of Scamandrus' (Astyanax's) abilities and those of his father, whom Plato names as a great defender of Troy. Therefore, the name Astyanax is the 'most true' name not only because it was bestowed by the gods but because it literally means Lord of the city (Plato, 1939, page 36).

Names are not mere labels, nor are they mere objects of subjective whimsy, all equal in applicability. They are caught up in myth, as seen above, and in rituals of naming, as in baptism. Myths and rituals of naming allude to original moments when objects obtain what is essentially 'theirs.' 'There is a first time for everything': this saying applies, with special force, to names. There was a first time that a particular person, place, or thing was called 'x.' Ancient Greeks resorted to gods to explain how certain things gained names—essentially through ostension (imagine the gods pointing at things on the earth and naming them, and men then taking their names from those things/places). Baptism is a ritual in which a name is assured legitimacy in the eyes of God, one who now is a monolithic and all-seeing being, beyond the multiple, distributed gods of the Axial Age (Bellah, 2011; Taylor, 2007).

While pristine (or baptismal, but here applied specifically to geographical) naming is ostensibly something that happens for the first time (as with human names for geographical things applied in colonial times), it is conceivable that animals have names for things, but are unable to express them in language. Animals would, so theorized, thus exhibit an extreme internalism to which the human animal is not subject and this, in the eyes of God, separates humans from the other animals, granting them special access to his [sic] favors in the bestowal and continuity of name-giving practices such as baptism, defined as "a religious rite symbolizing admission to the Christian Church, involving sprinkling the forehead with water or total immersion and generally accompanied by naming" (*Canadian Oxford Dictionary*, 2004, page 109).

As we will see below in the case of holy wells in Wales, water is indeed often bestowed with supernatural powers. In a geographical sense, water is also highly mobile and was used, where the water was still, to 'pointillize' its ritual use; in places where water was mobile, as in a river, it could serve as a boundary, but also as an ever-replenishing source of miracles and access to both magical and godly powers.

Humboldt's volcanic eye

In *Views of the Cordilleras and Monuments of the Indigenous Peoples of the Americas* (hereafter, *Views*), Humboldt (2012) identifies twenty-six geographical names of significance in myth and everyday life for indigenous peoples of South America. The entire work can be viewed as a gazetteer of place names as it contains a lengthy index of toponyms. A gazetteer is defined by Hill (2006, page 228) as "a dictionary or list of geographic names (placenames), together with their geographic locations, their feature types (e.g. lakes), and other descriptive information."

Views fits this definition of gazetteer, though it has not traditionally been viewed in this way. Instead, the critical edition with an introduction by Kutzinski and Ette (2012) constructs *Views* as a rhizomatic atlas of sorts (Daston and Galison, 2007). While this framing of *Views* as rhizomatic is valid, I will argue here that it is more properly seen as a gazetteer, but one with an amount of descriptive information that goes well beyond what most gazetteers provide

(Hill, 2006). *Views* is a gazetteer par excellence, an example of a tool forming part of an NTN for tracking indigenous mythical names through time. As such, it is also a tool for promoting continuity of both the proper (geographical) names and attached descriptive information. In a Kripkean sense, gazetteers are key parts in chains of communication that link past senses of proper names to present ones, with a key link between subsequent or overlapping names being the common referent (object in the world). The survival of geographical names such as those included in *Views* opens worlds of possibility to the viewer, just as the landscape features themselves opened the doors to mythical worlds for the original inhabitants of the life-worlds depicted (i.e. the indigenous peoples alluded to by Humboldt in his title). A link can also be made in (Wittgensteinean) terms of place-naming practices or belief systems providing bridges between the world 'out there' and the uttered, gestured, or written (i.e. performed) names stored in the bodies and brains in which and by which their performance takes place (Eades, 2015). We look more at this performed aspect in the next chapter.

The twenty-six toponyms of focus here all correspond to volcanoes mentioned in *Views*. A key distinguishing feature of the *Views* is the way it represents (or describes) named features in landscapes visually. We therefore select examples of volcanoes according to Humboldt's (2012, pages 126–127) visual typology. Three kinds of volcano shape are identified: a) still active, b) fallen in, and c) round peak. Examples of each kind of peak follow through descriptions by Humboldt, linked rhizomatically to each other through the inclusion of geographical names in *Views*. This is a rather unorthodox but highly effective way of organizing a gazetteer. It is effective because it allows Humboldt to build a narrative that deepens senses in which the features described hold meaning for indigenous peoples in the Americas (in this case South America).

As an example of a round-peaked volcano, we have "Chimborazo viewed from the Tapia Plateau" after a heavy snowfall on June 24, 1802 (Humboldt, 2012, page 224). Kutzinski and Ette (2012) note that this volcano was close to Humboldt's heart and that, for Humboldt, "volcanoes were at once the epitome of aestheticized nature and key features of a distinct cultural landscape" (Kutzinski and Ette, 2012, page xxiv).

Humboldt notes several important aspects of Chimborazo, including its elevation, both absolute and in relation to surrounding mountains and other large peaks globally. Plant and animal life is described in detail in terms of both elevation zones and the livelihoods made possible by their presence for local 'Indian' populations. Native legends indicate, according to Humboldt, that "one peak from the eastern crest of the Andes, which today is called the Altar (*el Altar*) and which partially collapsed in the fifteenth century, was once taller than Chimborazo" (Humboldt, 2012, page 226). El Altar is mentioned elsewhere in *Views* as an example of a collapsed volcano, but no visual representation is included, only textual descriptions and allusions under the headings for other features or place names. A temporal NTN is produced both by the indigenous narrative describing El Altar's collapse and its relation

along the Tapia plain to other volcanoes; and this narrative is both reproduced and added to, with scientific observation and narrative fragments from
Humboldt himself. The geographical name is tracked across space through
relation to other names; and across time through myth and legend.

Indeed, another reproduction of Chimborazo appears on page 123 of
Views, this time alongside the volcano named Carihuairazo. Much less an
outline and more an oblique-view relief map, "View of Chimborazo and
Carihuairazo" shows this favorite volcano of Humboldt's in monochrome
(Humboldt, 2012, page 123). It is accompanied by a geographical overview of
the entire Cordillera of the Andes, serving in metonymic relation for scientific
exploration of a part–whole relationship. As always, Humboldt takes a holistic
approach, including both fragmentary and seemingly tangential information
(such as extensive comparisons to heights of European mountains). At one
point he even compares Chimborazo to the work of Michelangelo, presumably
because of its airy, sublime, and almost cloud-like appearance (Humboldt,
2012, page 127)

Compare this representation to that of a still active, high-peaked volcano,
Cotopaxi, and part of its accompanying description:

> the shape of Cotopaxi is the most beautiful and the most regular of all
> the colossal peaks in the upper Andes. It is a perfect cone that, cloaked in
> an enormous layer of snow, shines dazzlingly at sunset and stands out
> delightfully against the azure sky.
>
> (Humboldt, 2012, page 64)

We again see seemingly extraneous aesthetic detail being brought to bear
upon material included in an ostensibly scientific text. Adjective upon adjective describing colour, shape, and beauty are used as a way of building a
narrative that incorporates quantitative detail as easily as it does the aesthetic
and native narratives. The remarkable shape of Cotopaxi, drawn on page 63
of *Views*, gives two views, one above the other. The top view is a painting
while the bottom view is a mere sketch. We see, above, a truncated white
triangle emitting smoke indicating its power and active nature. Two bands
show vegetated zones, and a small peak sits beside and below the main cone,
a jagged and

> small rock mass on the southwest side, half-hidden under the snow and
> spiked with points, which the natives call the Head of the Inca. The
> origin of this bizarre moniker is uncertain. According to a folk belief that
> exists in the country, this isolated boulder was once part of Cotopaxi's
> summit. The Indians claim that when the volcano first erupted, it hurled
> forth a rocky mass that had covered the enormous cavity containing the
> underground fire like the calotte of a dome. Some insist that this extra-
> ordinary catastrophe took place a short time after the Inca Tupac
> Yupanqui had invaded the kingdom of Quito and that the rock piece that

one can make out on the tenth Plate to the left of the volcano is called the Head of the Inca because its fall was the sinister omen of the conqueror's death.

(Humboldt, 2012, page 65)

Humboldt brings in native mythology and stories to bolster his own observations. He takes the indigenous view seriously, treating it with the objective eye of the anthropologist/geographer. The indigenous view is brought into *Views* in a way that was well ahead of its time, inspiring the likes of Darwin and others. Humboldt's evolutionary methodology *avant la lettre* created gazetteers as parts of atlases of knowledge and NTNs operating rhizomatically and often in fragmented fashion. This did not detract one iota from their effectiveness in spatial knowledge transmission and communication across vast stretches of time and space.

Beating the bounds

For hundreds of years in England, before boundaries were written into charters as secular means of calculating value, the bounds of the parish were overlain with rituals of spiritual significance (Pounds, 2000, page 67). There was a strong intergenerational aspect to the learning by the young of parish boundaries, and a need to pass on knowledge that, in medieval times, was not written down anywhere, and existed as a yearly performance corresponding roughly to the ancient Roman Rogation days. According to Pounds (2000, page 76),

> parishioners had to know the bounds of their own territory, however tortuous and obscure they may have been. The 'beating' or perambulation of the parochial limits was thus a ritual of considerable antiquity and great importance. It served to acquaint the young with the extent of the parish and to refresh the memory of the old.

Pounds notes the secular quality of 'beating the bounds' and the fact that knowledge of boundaries and maintenance of paths and walking trails with which they corresponded would have a positive effect on the image of the parish. Those parishes that collectively knew their boundaries were seen in a better light than those that let them slide into oblivion through loss of collective memory of the spatial limits of the territory associated with the local church. There was also a physical, performative aspect to the rogations (as we will refer here to the beating the bounds rituals), notably that children would be touched, 'bonked,' or beaten with objects corresponding to key points or 'nodes' on the perambulation's path. So, for example, a tree or boulder lying at a crucial turning point in the rogation path would find itself ripe for 'bonking' a child's head, or a branch could be used to touch the child at that key point in the path, adding a physical reminder to the visual stimuli encountered along the way (Kate Distin, personal communication). Bodily performance was thought (and indeed demonstrated) to aid in intergenerational memorization of the spaces and boundaries of the parish and thus, of its areas of (calculable) value.

The whole thing, the perambulation, physical contact with 'nodes' in the parish boundaries, and the fact that the whole boundary had to be walked to be considered valid (Pounds, 2000), before a time when the boundaries were inscribed or written down, had something of the flavor of 'contagious magic' (Thomas F. Thornton, personal communication), whereby the learning of boundaries and key points was somehow transferred to the younger generation through a kind of spontaneous transference facilitated by contact with a part of the boundary (Mauss, 2001). The physical performance and transference facilitated the creation of intergenerational memes, or 'units' of culture (cultural objects), for transmitting the nodes and by extension boundaries of the parish (Distin, 2005).

Pounds elaborates upon the antiquity, and thus intergenerational nature, of the rogation: "the perambulation of the parish was essentially a secular ceremony, but custom had overlaid it with religious symbolism and had conflated it with the ritual of Rogation Day. The latter itself was a ceremony of confusing antiquity."

Of more importance, however, to the geography of names is the fact that the boundaries so perambulated were also named. As described above, there were 'nodes' or points in the perambulated path. These nodes could be trees, stones, wells, or crossroads. But as a GIS or NTN, in this case it was one that was embodied intergenerationally in the brains and bodies of those walking the boundaries, and a more complex set of spatial 'primitives' (to use GIS parlance) was at play (Collignon, 2006). A boundary could thus consist of defined linear features such as roads, rivers, or pathways. Areal features that provided constraints for the definition of parochial limits included adjacent parishes, fields, or manorial territories (Pounds, 2000). Points (called nodes above), lines, and areas are the so-called 'graphic primitives' of modern-day GIS' vector data models (Schuurman, 2004). I am not claiming that GIS is somehow derived from rogation. I instead suggest that rogations corroborated claims for particular (predominantly vector-based) ways of structuring and modeling (named) geographical space as opposed to other (e.g. raster-based) ways of modelling the same space, less conducive to naming practices.

For our purposes we move forward several hundred years to modern-day rogation practices that keep alive the so-called 'beating of the bounds' rituals. These can be found across England and are perhaps now imbued with more of a carnival spirit (Hutton, 1994). An Oxford or Camden rogation in the twentieth century might involve religious figures, but it need not do so. Stops at pubs, singing, and photography would not be out of place on a 'beating the bounds' outing, nor were they before (with the exception of photography), but the modern-day equivalent would seem to be a curious mix of post-secular fun and an excuse to get some fresh air and exercise away from the office as much as anything else. We therefore turn to two primary documents describing modern-day rogations. These are "St. Michael at the North Gate, Oxford: Beating the Bounds" (Martin, 1961) and "Beating the Bounds of Camden: A Long Distance Walk" (Holmes, 1993), found at the Bodleian Library in Oxford.

We explore these two pamphlets for evidence of 'tools' such as names used to produce new kinds of boundaries, selves, and senses of space in keeping with modern sensibilities. The parish no longer structures everyday lives in secular or religious senses. Property is surveyed using precision instrumentation without need for human-embodied and performed GIS (McGrath and Sebert, 1999). But modern-day rogations may be more part of what can qualify as movements to quantify selves through measurement of distance as an amount of exercise taken and thus a milestone achieved in, for example, losing weight or gaining fitness. These are the secular grails I would posit as sometimes now driving modern rogation practices, in addition to continuities associated with celebration and general merriment (Ifould, 2013; Hutton, 1994; Holmes, 1993; Martin, 1961).

It is useful at this point to pause to consider the definition of names used in this book, to establish their usefulness and their validity in the examination of rogation practices. Remember that we define names in a Fregean-descriptive sense as noun phrases that refer to particular things (Hawthorne and Manley, 2012; Moore, 1993, page 1; Bach, 1987; Evans, 1982); the second 'prong' of our definition is the sense in which names are 'proper' as explored by Kripke (1981). Noun phrases are descriptive while proper names are not (at least not straightforwardly so). The latter obtain meaning and reference in an a posteriori sense, i.e. over time and in retrospect (Berger, 2011; Fitch, 2004; Kripke, 1981). Rogation practices use both noun phrases (descriptions of particular things in the world) and proper names as tools/practices for making sense of territorial space through time. There is no contradiction in asserting both Fregean and Kripkean senses of names, and we seek here, in line with new currents in philosophy (Hawthorne and Manley, 2012), to overcome a dichotomy in the literature between the two through their application to space. This adds the all-important modifier, space (i.e. geography), to names for exploration of the *geography* of names, a category notably left out of most of the philosophical literature. To date, in philosophy of language works, for example, only a very few have looked explicitly at spatial referencing (see Evans, 1982, especially pages 278–284). In geography, spatial ontologies are beginning to gain intellectual ground, and the present work places itself firmly within this sub-discipline of geography and geographic information science (Mark et al., 2011).

The tools and ritual practices of rogation form parts of NTNs that make sense of, and communicate ideas about, space through time (i.e. intergenerationally). Winchester (1990, page 3) discusses some of these ideas in relation to the basic spatial unit that is named in rogation, namely, the parish:

> the network of parish boundaries formed an invisible web which both bound families into communities and divided communities from one

another. The landscape of local administration was both ecclesiastical (determining the church in which a person was baptised and buried and to which he paid tithes and other dues) and civil (dictating the official to whom he was responsible for payment of taxes and rates, for example). The boundaries of parishes and other units of local administration mattered greatly to our ancestors. Before the local government reforms of the nineteenth century, the parochial basis of poor relief and many charities and schools gave considerable importance to the parish in which a person was born and thus endowed parish boundaries with, perhaps, undue significance.

Thus, the parish was many things and it was caught up in political, commercial, and governmental circuits of power and mentalities. Defining the precise boundaries corresponding to parish names was thus a process that was conflictual and contested at all stages.

Relevant to power and conflict are units of language smaller than the noun phrase or proper name (by which we here define geographical names), namely place-name elements that occur as parts of place names for defining or alluding to parish boundaries (Winchester, 1990, pages 86–87). A selection of these includes several referring to conflict, division, or disagreement. For example, *calenge* (from Middle English) means challenge or dispute, as seen in Worcestershire's Callans Wood; *ceast* (from Old English) means strife or contention as in Wiltshire's Chesland or Essex's Chest Wood; flit (from Old English) means strife or dispute, as in Sussex's Flitteridge (Winchester, 1990, page 86). Winchester's (1990, pages 86–87) glossary of "Place Names Recording Boundaries" includes, for the most part, word particles referring specifically to boundaries or divisions, with seventeen boundary-specific word parts; and one (ecg, from the Old English) that refers to topographical edges such as cliffs or hill scarps that might have been used for boundaries.

And how, precisely, does the beating of a boundary marker with a stick (without, for the sake of argument utterance of a name) constitute naming in the senses used here? If the parishioner participates in rogation but beats the boundary without utterance, can the use of names as tools be said to exist? It can, if one considers that reference (naming) can occur in thought, demonstratively (Bach, 1987), or as part of mental and cognitive maps (Evans, 1982, pages 151–204; O'Keefe and Nadel, 1978). The latter are embodied in brain processes that are posited to occur simultaneous to the beating of the bounds with sticks, canes, branches, or other objects. There is, in essence, a set of 'neural correlates' to the spatial practices of beating the bounds. These correlates can be said to be both individual and shared across individuals in the sense that these 'brain maps' resemble each other in specific ways. We explore these and related concepts in the next chapter (4), on neurogeographies and naming.

Returning to modern-day Oxford and Camden rogation practices, we can see that despite the secularization of rogation, from the perspective of names much remains the same. The urgency of transmitting information about parish boundaries was once quite pressing for priests and their parishioners, the impetus being the collection of taxes based on (i.e. the demonstration of value based on areal) extent and use of parish lands (Pounds, 2000). The children would in due course be responsible for such collections. In the modern world, with that urgency (i.e. of tax collection, or more generally demonstration of value) firmly in the hands of government (for which the parish priest performed a sort of proto-governmental operation), rogations are a chance to not only commemorate the past and its landscapes but also to create a positive, structured, image of the local (Leslie, 2006). This is accomplished, as in the past, through names and naming, but its manifestation now is somewhat altered or evolved and the commemorative aspect looks more firmly to the past, with less anxiety around transmitting boundaries to future generations. There is less riding on it now that taxes and curbing criminal behavior through control of space and territory are firmly in state hands.

Names in modern rogation practices still track parish extents. When beating the bounds of Camden (Holmes, 1993), that name is being tracked, and the performance, maps, and utterances forming part of the rogation are, in turn, part of the NTN for producing correspondence between 'Camden' and its spatial extent. Names, in this sense, are precisely about reducing ambiguity between what is, and what is not, Camden (Malt, 2015), and the practice is also about identity. One can imagine walking through Hampstead Heath to the height of land, looking over and realizing that beyond that road (at the top of the hill) one is no longer in Camden. One is thus naming by reducing the options for moving in space. This could also be referred to as a 'creative constraint' in modern psycho-geographical parlance, whether walking through progressive Camden or relatively staid Oxford. The latter contains a plethora of urban medieval walls, streets, alleyways, and indeed boundaries that seem not to heed natural and human structures. The naming aspect remains the same, however, as when the parishioners beat the bounds of the parish named 'St. Michael at the North Gate' (Martin, 1961). This first level of naming is that of the 'proper' (geographical) name (Kripke, 1981), which is quite distinct from, and is used here in addition to, the Fregean descriptive sense of name (as noun phrase) described below.

The second level of naming corresponds to 'places along the way' of the rogation. These are the "singular noun-phrases that are used to refer to particular things" (Moore, 1993, page 1). For example, there is an 'X' mark on a stone on the north side of the Bodleian Library (Martin, 1961, page 13). The X mark is fixed to a permanent feature of the landscape (a foundation stone), which has the psychological effect of reducing ambiguity about the extent of 'St. Michael at the North Gate'. (It also increases ambiguity because the boundary thus defined cuts the Bodleian Library itself in half). However, as Bonaventura (2007, page 117) has pointed out, "boundary features can offer a sense of physical permanence in their primordial boulders, ancient trees or craggy shorelines, the bounds of property are wholly the transient handiwork of the human mind." This is as true in Oxford, and perhaps even more so, given the transience of human landscapes in relation to natural surroundings, as in the rural non-city settings described by Bonaventura in his exploration of how American settlers attempted to transplant rogation practices in the new world to reduce ambiguity in radically new surroundings (Bonaventura, 2007).

Modern-day rogation practices benefit from maps to an extent that was not possible in medieval or even post-medieval times. Though boundary books were used to inscribe representations of boundary markers for later potential dispute resolution (Holmes, 1993, page 3), pre-modern rogations relied entirely upon mental and cognitive maps for the storage and representation of spatial forms (Eades, 2015). I would argue that this primary function, that of storing mental maps for intergenerational knowledge transmission, separates pre-modern from modern and post-modern rogation practices. In Martin (1961, page 8), for example, we have a map of modern-day Oxford with a drawing of the boundary and 'x marks the spot'-style points (nodes) showing stones themselves inscribed with Xs for beating with rods, canes, or willow branches. The map is a central part of all modern hunts for parish boundaries, whether in church-produced pamphlets (Martin, 1961; Holmes, 1993) or through the use of Ordnance Survey maps that Winchester (1990, page 35) recommends as the starting point for explorers and ramblers seeking to 'recover' lost boundary stones for (re)mapping into the landscape, into their 'personal files' of *de re* belief, and for metaphorically finding one's way (Bach, 1987). The maps thus serve a performative function such that their inscriptions can serve as repositories of later actions (performances) upon which the maps are, in turn, further updated, in a virtuous cycle of performance and inscription (Eades, 2015). Belief in 'the deed' is indeed very important to modern-day ramblers and rogationists, as there are no respected 'armchair' variant sub-cultures, hypothetical travelers whose belief in description ('the word', or *de dicto* belief) would outstrip a correlated emphasis on 'the deed' (or *de re* belief in Bach, 1987). 'Armchair' and 'real' travelers are, it would seem, species apart. We can hypothesize or speculate (but not prove) that in past times the separation may not have been so extreme. The next chapter should bring some clarity to this question.

Important aspects of rogation practices both ancient and modern include *ritual, belief* (including magic), *continuity*, and *colonization*; and they serve several simultaneous purposes, many of which are now obsolete. Each of these aspects is explored below by way of summarizing rogation practices in relation to geographical names. In terms of *ritual*, rogation practices were always done in the same way, e.g. clockwise, with a full 'rotation' of the entire boundary, regardless of how long it took. Rogations also occurred very close to Ascension Day, either on the day itself or during the previous week e.g. on a Sunday. They were also timed to correspond with the rogations of adjacent (neighboring) parishes. The ritual aspect of rogation meant that it was rigid in terms of method and timing, but this was in keeping with its proto-governmental function of calculating (albeit fairly qualitatively) boundaries precisely so that accurate collection of taxes could take place. This, in turn, contributed to the capture of criminals and to charity for the poor, both of which parish churches were responsible for (Pounds, 2000; Reff, 2005).

In terms of *belief* in the practices carried out during rogations and the magic associated with that belief, the power held by the church meant that parishioners believed they were acting in the name of God by participating in beating the bounds ceremonies. The extensions to their bodies in the form of sticks, wands, and canes meant that participants believed they were extending the continuity of the boundaries through time both objectively (i.e. in the eyes of the world and of God) through the use of such implements and subjectively by using the same implements or boundary objects themselves to 'beat' children and other participants into remembering object-positions (i.e. of nodes, linear features, and areas), thus adding to the security of the boundary being mapped through the rogation. Names encountered corresponding to referent objects were seen to stand in for the objects themselves, either as descriptive noun phrases or as proper names. A modern-day boundary-producing instruction such as "walk past the Gatehouse Pub and turn left onto Hampstead Lane" (Holmes, 1993, page 10) could as easily have served as a medieval as a modern boundary definition (and could well have induced a need for a morale-boosting refreshment stop), but its easy-to-remember land marking occurs precisely through such naming practices. Usually some kind of marker would have been placed at a 'corner' location such as this, marked by the priest with a code and then beaten by the participants. The rods serve the magical function of transforming the object in the world into something with a life of its own, and also giving it permanence not justified by the physical permanence of the object, and more to do with anxiety about their impermanence. Thus, the need to conjure up (but nevertheless very real) *continuity* through the 'contagious magic' of the wands, canes, and child beatings for inducing intergenerational knowledge transmission and enduring legacy of the names.

Colonization will also be explored below, under holy wells of Wales, being caught up with the quasi-colonizing territorial logic of defining boundaries to their fullest extent, without overlap or empty space between parishes (Elden, 2013). As noted by Winchester (1990), landscapes in England varied between northern and southern locales but basically conformed to township, vill, and (ecclesiastical or civic) parish structures. Manor areas and church parish boundaries were codified first through performances on the land in a pre-inscriptive early Christian culture that attempted to replace all vestiges of pagan and Celtic 'cult' practices. The locations of churches, for example, often promoted continuity of the *sites* of those older practices while reforming older beliefs in conformity with newer religious and monotheistic Christian structures. Churches were placed to clash with or replace assemblages of holy trees, megaliths, druidical stone circles, holy wells, and the like, without displacing the actual location of the place of worship (Jones, 1954). Contemporaneous to boundary-making practitioners' (priests and parishioners) increasingly in sophisticated abilities to create and consult maps and other inscriptions post-medieval times, colonization became complete, first in England and later in Wales, though both retain place names associated with sites from much earlier times (and especially in Wales, from pre-Christian times). The durability of toponyms extends their life well beyond the death of the cultures from which they were produced (Cunliffe, 2013).

As has already been noted, continuity was fostered by the inclusion not just of children but of a cross-section of differently aged parishioners. Children were included as a kind of insurance policy for collective-memory preservation, to the point where they could be traumatized into remembering the boundaries, lest their precise locations be lost to future generations (with attendant loss of territory, prestige, and tax revenues). Modern-day practices are not innocent of such abusive practices. Martin (1961, page 16) notes that (with reference to Oxford), at the same time as a cheer is thrown up at the conclusion to the rogation, leading to a scramble for coins tossed in the air,

> for many years it has been a custom among under-graduates of the College to heat pennies on a shovel over the fire and to throw them, mingled with the official ones, and to observe the conflict of avarice and apprehension among the young recipients of their bounty. In 1938 a generous and kind-hearted American Curate, warned of this sadistic English custom, provided each boy with a small pair of tweezers, with the disappointing result that any boy who paused to use them got little, while his more valiant comrades obtained 'the Lion's share'.

This serves to demonstrate that, as with earlier practices, some of the children themselves later become knowledgeable elders forming key components in NTNs for ensuring continuity of not only the boundaries but also the rituals and practices (such as that described by Martin) associated with their propagation. It is important to remember that the intergenerational component served both colonizing ends, as new names corresponding to Christian practices became embedded not only in the brains and bodies of the participants but through the 'beating' in the land (and again) in the bodies in turn, and also to propagate features much older than Christianity that had been appropriated by the latter according to its needs or simply opportunistically, as it would be easier to appropriate a well or stone circle site than seek to replace it. Older features so appropriated by Christian rogation practices would effectively lose their older, cultic and magical, power through this appropriation. But conservative forces, and materiality, ensure that the older names often survived, especially in areas marginal to mainstream England such as Wales, Scotland, or Ireland.

Thus, to reiterate, several purposes were served by rogations. First, they delineated property for taxation purposes. Second, as seen by the beatings and "sadistic English custom" (Martin, 1961, page 16), one of the main purposes of rogation was discipline, both of bodies and of the land. Third, they transmitted knowledge across generational boundaries. Last, they served "to supplicate for Divine blessing on the crops" (Holmes, 1993, page 3) in traditional times. It is to those traditional times that we now turn in an examination of one particular kind of feature in the landscape of belief, that of water—or to be more exact, the belief in its holiness, resulting in the creation of holy wells, often at boundaries, and more often than not with very interesting and informative names indeed.

Water, wells, and Wales

Jones (1954, page 29) notes, with respect to belief, rites, and superstitions associated with or deriving from well-cults that,

> [w]hen we dismiss the fictitious tales and romantic inventions as unhistorical we must accept the fact that such tales and such inventions influenced the ways of thought of our ancestors. So it is with the well-cult. We are not concerned so much in this study with the element of truth in the traditions as we are with the undisputed effect the well-cult had on the human mind.

Jones' (1954) classic on the holy wells of Wales is a study of belief in the power of these wells to heal the sick. Since antiquity, wells were associated with megaliths, sacred trees, and other sites where power was believed to reside—most importantly, where water was seen to emerge from the ground from springs, pools, or human-made wells for accessing ground water. An association with water can be found in the primary symbol of Wales, the red dragon that appears on its flag, as we will see below.

The Princes of Gwynedd Guidebook (Cyngor Gwynedd a Phartneriaid, 2013, page 22) tells us that at Dinas Emrys,

> Vortigern, a powerful ruler of Britain in the 5th century AD, was rallying against Anglo-Saxon invaders. He sought to build a stronghold on the strategic hill. Each day he set his builders to work, and each morning they woke to find their work undone. Piles of rubble lay where newly built walls had stood. The magicians who advised Vortigern suggested a solution: he must sacrifice 'a fatherless boy' and sprinkle his blood upon the site where he wished to build … . An appropriate boy was quickly found. Before his blood could be shed, however, the boy convinced Vortigern that the problem lay with the site he had chosen. Beneath the hill, he explained, was an underground lake containing two dragons. Digging proved the boy to be right, and when they were released the two dragons— one red, one white—fought each other. The red dragon was eventually victorious, and this symbolic defeat of the Saxons by the native British people is commemorated on the Welsh flag to this day … . Vortigern named the castle which he finally built after the boy who had advised him. He was called Myrddin Emrys—but today we know him as Merlin, the great and wise adviser of King Arthur.

The connection of the power of the site with water is explicit. The two dragons are said to have emerged from an underground lake, an unseen source of water emerging from beneath the earth in the form of two dragons. The guidebook goes on to posit a question with respect to the 'water area':

> Today visitors to Dinas Emrys can see the ramparts of the ancient hillfort, the stone foundations of the medieval tower and the silted pool below. Excavation showed the pool was once a timber-lined cistern; could there be a link between this watery area and the legendary dragons' lake?
>
> (Cyngor Gwynedd a Phartneriaid, 2013, page 22)

Dinas Emrys is a site of magical import in local myth and in the continuity of Welsh identity. The name is evocative, therefore, not only of Wales (or Cymru) but of Merlin and Vortigern, two powerful early figures in Welsh myth. It has the power to bring to mind a symbol of Welsh origin and identity through the evocation of the red dragon that symbolizes victory of the natives over the Saxons and thus the power of local inhabitants (i.e. the indigenous people) over outsiders. Its association with water is inextricably linked to other forms of water-worship in Wales, as described so well by Jones, in which, "[i]n ancient and pre-Christian times wells, springs, lakes and rivers were worshipped as gods or regarded as the abodes of gods, and associated with them were purification ceremonies, sacrifice, divination, fertility, healing and weather charms" (Jones, 1954, page 12). What better symbol of healing of Welsh woes than the saving of the land from outside (Saxon) invaders?

In 2014 I had a chance to visit Dinas Emrys and see the cistern remains from which the dragons were said to have emerged. The site, on the day of my visit, was exceptionally tranquil and I felt moved to have the parking space to myself at a site of such profound significance. Before arriving at the site it had seemed to me that the place of the red dragon that we see today on the flag of Wales would be cause for celebration on a par with parliament buildings or the changing of the guard (i.e. crowded and spectacular). Instead, I found the cistern atop the hill where Merlin's Castle was once said to sit, with not a soul in sight other than myself and a research assistant companion. A medium-sized tree grows fairly close the square-shaped sunken enclosure, the latter the remains of the cistern referred to in the Princes of Gwynedd Guidebook. From this spot one can take in a view of the road, river, and forestlands below. It is moving to contemplate the origin of not only the source of one of the most powerful myths of Gwynedd-land, the ancient and hermetic home of princes in what is now northwest Wales, but of the country as a whole. Dinas Emrys is thus a sort of 'spring-loaded' proper name, or place meme, holding a great deal of associated knowledge accumulated over generations.

Dinas Emrys has a key place in the Welsh NTN for tracking knowledge systems associated with identity, territory, belief, and boundaries. The extent of the entity (Wales) referred to by 'Dinas Emrys' is geographically complex, bounded by coastline on three sides and a wall on the other (the latter defining what is and is not Wales). A sort of 'spiritual center' could therefore be said to reside at this 'watery area' on a hill in a quiet valley in Snowdonia. Modern Wales has clearly moved on, however, from well-worship and from what might be called pagan worship of watery places. The flag itself now stands for Welsh identity, and there's not much need to visit the actual site of which Merlin made so much. But it is instructive to look for sources of older belief, in this case in wells, for toponymic evidence of boundary-making. This evidence holds a key to unlocking how both proper names and noun phrases for referring to water form toolsets for making sense of the world.

Jones (1954) notes that several named places served dual religious/secular functions in property and parish boundaries. In practice it was very difficult to separate the holy aspect of a well from its function as a place marker. Jones (1954, page 56) notes that,

> A holy well in Burton parish (Denb.), now only a faint depression in a meadow, is of special interest since it was one of the fixed boundary marks between England and Wales, being described as such in a Minister's Account for 1448. The Three Counties Well near Llanymynech (Mont.) is so named because the borders of Shropshire, Denbighshire and Montgomeryshire meet at that spot. A few examples are found in deeds after the close of the Middle Ages, such as *Fynnon Sevte* a boundary of Ystradvellte parish (Breck.) in 1502, and in 1667 *Ffynnon Ryddwern* and *Ffynnon Varred* occur as boundaries of the manor of Neath Citra (Glam.) One such record shows the extension of the cult of Derfel Gadarn to South Wales. A survey in 1597–8, describes *Dervel's Well* as a boundary mark of the manor of Landimor in Gower. There was some argument about Moor Well on the boundary of Gresford parish (Denb.) In 1642, the curate of Gresford and some parishioners came to the Moor Well as they had done for many years before, to claim the well to be in Gresford parish. However they saw some soldiers by the well, 'which wanted to see their fashions', whereupon the curate and his company departed and never came to claim the well again.

Lists such as these abound in Jones (1954). The above quote serves two functions. The first is to note that the secular functions of boundaries that gave impetus to their inscription also, by that very inscription, served as part of the impetus of their religious aspect. The latter aspect was marked significantly by what Jones calls "The Well-cult in Medieval Wales." The writing down of locations of wells that served as boundary markers in charters, accounts, and deeds and indeed on maps was part and parcel of a newly emerging governmental paradigm and mentality that piggybacked proprietary logics onto prior parish-based oral performances. Such performances in religious life corresponded to a topography of belief that was in large part concerned with saints and their stories, and these in turn were based in large part upon the sacred status of water, springs, wells, and the like. The second aspect of note is, therefore, the residual religious character of the well boundary markers moving into post-medieval times. To quote Jones (1954, page 44) again:

> Some local saints, whose piety caused their memory to be revered in remote districts, but whose names never figure in menologies, are, nevertheless, remembered in the names of wells and chapel sites. The Cardiganshire Ebwen (perhaps Mebwen—wyn) is one of those forgotten saints. A royal confirmation to Strata Florida in 1426 recites a charter from Rhys ap Tewdwr in 1184, and among the boundary marks is *Finan Mebwyn*, between Castell Flemys and Coet Mawr, and it also mentions "the part of the nuns of *Fennaun Vebwynn.*" A deed, dated 1704, describing properties in Lampeter parish, as *Tyr Capell ffynnon Ebwen*, proves conclusively that here we are in the undoubted presence of a holy well with its attached chapel, named after a local saint. It is possible that Elin, whose name is recorded in Professor Rees' map in Ffynnon Elin, south of Llanilar (Card.), is another example, and also Govan and Degan (Pem.), and perhaps Gwenno (Carm.) and Mymbyr (Caern.)—all of whom had wells bearing their names.

As we see in the above two quotes, noun phrases including the word *Ffynnon* (and variations thereof) were often modified by the names of saints, either through the juxtaposition of the (proper) saint names as modifiers of the generic terms, or by their association with religious stories attached to the toponyms. The partial lists above are in turn parts of Jones' gazetteer of religious toponyms 'tagged' to water features in Wales. Like Humboldt, Jones has surveyed a topography of belief that is both fluid and hybrid, as described above. Part II of *The Holy Wells of Wales*, comprising around one third of the volume, is devoted to a more 'formal' gazetteer of list of 'ffynnon' place names, separated into five classes labeled A through E. Class A wells bear saint names and designate Drindod, Duw, Mil Beibion, and Pas (or, Trinity, God, Holy Innocents, and Easter, in that order); class B wells are associated with "churches, chapels, feasts, pilgrimages" (Jones, 1954, page 140); class C are healing wells lacking either saint or church names; class D are named after secular people or minor saints; and class E are miscellaneous. The point of reciting these lists is to note the rich and varied character of the kinds of names associated with an underlying (past) landscape just beneath the surface of secular life in Wales, hints of which are evident only in the landscape of names (Jones, 1954).

The lists and extended quotes from Jones serve as a backdrop for discussing how names evolve and continue through political, secular, and religious paradigm shifts. The material traces of these changes can be seen or evidenced sometimes directly in the landscape or in works that preserve the 'tags' and labels that once referred to (now lost) topographies of belief. There was (and is) a blurring of the secular and religious (as seen in the various classes of well names) and of the performed and inscribed nature of the names. Belief *de dicto* (by the word) and *de re* (by the deed) continues in manifold formats across all of these divides and makes them seem artificial. For place names can be inscribed on deeds, charters, tourist brochures, and maps; they can be uttered, lectured, or otherwise orated in churches and indeed, online in our modern era. These are the multiple and overlapping forms names now take, and their transmission is caught up in evolving and overlapping practices that form NTNs that communicate and sometimes change the underlying information.

Extending names into cognitive terrain

Names, like water, are everywhere. They are the stuff and scaffolding of human experience. Our names stand for us, and vice versa (Blount, 2015, page 616). Our bodies are mostly made of water, as we are made up of names, of things, of molecules, of our 'essential' selves. In religion, the autochthonous nature of water is linked to names for gods, to naming practices in baptismal rites, for example, and to pre-Christian belief in its healing power. The names of water are in us and all around us, as are the names of every other substance that can be imagined (Klein and Lefevre, 2007). We explore in the next chapter what I have come to refer to as 'computable names' (Hodges, 2014) in the sense that there are some names, especially geographical names and those that correspond to faces, that 'make sense' in certain core areas of the human brain. We process place and face names in the hippocampus, for example, in conjunction with other brain structures such as the amygdala (Quiroga, 2012). It is to these, and others, that we now turn.

References

Bach, Kent. 1987. *Thought and Reference.* Oxford: Oxford University Press.

Bellah, Robert N. 2011. *Religion in Human Evolution.* Cambridge, MA and London: The Belknap Press of Harvard University Press.

Berger, Alan (ed.). 2011. *Saul Kripke.* Cambridge: Cambridge University Press.

Blaut, J.M. 1979. Some Principles of Ethnogeography, in Gale, S. and Olsson, G. (eds). *Philosophy in Geography.* 1–8. Dordrecht: D. Reidel.

Blount, Benjamin. 2015. Personal Names, in Taylor, John R. (ed.). *The Oxford Handbook of The Word.* Oxford: Oxford University Press.

Bonaventura, Allegra di. 2007. Beating the Bounds: Property and Perambulation in Early New England. *Yale Journal of Law & the Humanities.* 19(2). 115–148.

Canadian Oxford Dictonary. 2004. Don Mills: Oxford University Press.

Collignon, Beatrice. 2006. *Knowing Places: Innuinnait, Landscapes and Environment.* Calgary: CCI Press.

Cunliffe, Barry. 2013. *Britain Begins.* Oxford: Oxford University Press.

Cyngor Gwynedd a Phartneriaid (Gwynedd Council and Partners). 2013. *Princes of Gwynedd Guidebook.* n.p.

Daston, Lorraine J. and Galison, Peter. 2007. *Objectivity.* Brooklyn, New York: Zone.

Distin, Kate. 2005. *The Selfish Meme.* Cambridge: Cambridge University Press.

Distin, Kate. 2013. Personal Communication. Oxford.

Eades, Gwilym. 2015. *Maps and Memes: Redrawing Culture, Place, and Identity in Indigenous Communities.* Montreal and Kingston: McGill-Queen's University Press.

Elden, Stuart. 2013. *The Birth of Territory.* Chicago: University of Chicago Press.

Evans, Gareth, 1982. *The Varieties of Reference.* Oxford: Oxford University Press.

Fitch, G.W. 2004. *Saul Kripke.* Chesham: Acumen.

Hanna, Patricia and Harrison, Bernard. 2004. *Word & World: Practice and the Foundations of Language.* Cambridge: Cambridge University Press.

Hawthorne, John and Manley, David. 2012. *The Reference Book.* Oxford: Oxford University Press.

Hill, Linda L. 2006. *Georeferencing: The Geographic Associations of Information.* Cambridge, MA: MIT Press.

Hodges, Andrew. 2014. *Alan Turing: The Enigma.* London: Vintage.

Holmes, Malcolm J. 1993. *Beating the Bounds of Camden: A Long Distance Walk.* Camden: Leisure Services Department.

Homer. 2008. *The Iliad.* Oxford: Oxford World's Classics.

Humboldt, Alexander von. 2012. *Views of the Cordilleras and Monuments of the Indigenous Peoples of the Americas.* Chicago: University of Chicago Press.

Hutton, Ronald. 1994. *The Rise and Fall of Merry England: The Ritual Year 1400–1700.* Oxford: Oxford University Press.

Ifould, Rosie. 2013. Do the Maths. *The Guardian Weekend.* 16 November.

Jones, Frances. 1954. *The Holy Wells of Wales.* Cardiff: University of Wales Press.

Klein, Ursula and Lefevre, Wolfgang. 2007. *Materials in Eighteenth-Century Sciences: A Historical Ontology.* Cambridge, MA: MIT Press.

Kripke, Saul. 1981. *Naming and Necessity.* Malden: Blackwell.

Kutzinski, Vera M. and Ette, Ottmar. 2012. The Art of Science: Alexander von Humboldt's Views of the Cultures of the World, in Humboldt, Alexander von. *Views of the Cordilleras and Monuments of the Indigenous Peoples of the Americas.* xv–xxxv. Chicago: University of Chicago Press.

Leslie, Kim. 2006. *A Sense of Place: West Sussex Parish Maps.* Bognor Regis: West Sussex County Council.

Malt, Barbara C. 2015. Words as Names for Objects, Actions, Relations, and Properties, in Taylor, John R. (ed.). *The Oxford Handbook of the Word.* Oxford: Oxford University Press.

Mark, David; Turk, Andrew, Burenhult, Niclas, and Stea, David (eds). 2011. *Landscape in Language: Transdisciplinary Perspectives.* Amsterdam: John Benjamins.

Martin, R.R. 1961. St Michael at the North Gate. Booklet. Oxford.

Mauss, Marcel. 2001. *A General Theory of Magic.* Abingdon and New York: Routledge.

McGrath, Gerald and Sebert, Louis. 1999. *Mapping a Northern Land: The Survey of Canada, 1947–1994.* Montreal and Kingston: McGill-Queen's University Press.

Moore, A.W. (ed.). 1993. *Meaning and Reference.* Oxford: Oxford University Press.

Nagel, Thomas. 1986. *The View From Nowhere.* Oxford: Oxford University Press.

O'Keefe, John and Nadel, Lynn. 1978. *The Hippocampus as a Cognitive Map.* Oxford: Clarendon Press.

Plato. 1939. *Cratylus.* Cambridge, MA and London: Harvard University Press.

Pounds, N.J.G. 2000. *A History of the English Parish: The Culture of Religion from Augustine to Victoria.* Cambridge: Cambridge University Press.

Quiroga, Rodrigo Quian. 2012. *Borges and Memory: Encounters With the Human Brain*. Cambridge, MA: MIT Press.

Reff, Daniel T. 2005. *Plagues, Priests, and Demons: Sacred Narratives and the Rise of Christianity in the Old World and the New*. Cambridge: Cambridge University Press.

Schuurman, Nadine. 2004. *GIS: A Short Introduction*. Malden, MA: Blackwell.

Stuart, David. 1994. *Classic Maya Place Names*. Washington, DC: Dumbarton Oaks Trustees for Harvard University.

Taylor, Charles. 2007. *A Secular Age*. Cambridge, MA and London: The Belknap Press of Harvard University Press.

Thornton, Thomas F. 2013. Personal Communication. Oxford.

Winchester, Angus. 1990. *Discovering Parish Boundaries*. Haverfordwest: Shire Publications.

Wittgenstein, Ludwig. 2001. *Tractatus Logico-Philosophicus*. Abingdon and New York: Routledge Classics.

Wittgenstein, Ludwig. 2009. *Philosophical Investigations*, 4th edition. Chichester: Wiley-Blackwell.

4 The neurogeography of names

Names refer and are referred to; names are both reference and referent, in a complex interplay of inscription, performance, embodiment, and transmission, as explored in the previous chapters. The latter (transmission) has not yet been fully explored, but as demonstrated below, place names are, or form parts of, memes for transmitting spatial information across time and space. Intergenerational spatial knowledge constructs are passed along from generation to generation through performances, utterances, and inscriptions of place names. These form part of what I call place memes (Eades, 2015) for communicating names (Kripke, 1981) and named structures such as journeys, commemorative visits to places, and stories. Names and their performance or storage in bodies and brains and external media such as maps become parts of NTNs that are, in turn, generative of place-naming practices (Hanna and Harrison, 2004; Wittgenstein, 2009). These practices form bridges between place names and their referents in the real world. Thus, there is two-way traffic between word and world that is bridged by practices. Names themselves form these bridges between language and worlds, and these names (both proper, and noun phrases) allow for structures of place to be built up in and through the cultures in which the practices are situated, fostered, and adapted to variable exogenous forces such as those associated with colonization.

Thus, naming practices vary alongside variations in human culture. But this does not diminish the ubiquity of names. Place is literally a ubiquitous phenomenon, and this fact is reflected in the ubiquity of names (as noted at the end of the last chapter). Even non-places (Auge, 2009; Cowen, 2015) can be given names (e.g. airport lounge or fast-food restaurant) and thus form parts of NTNs and place-naming practices in cultures around the world. All cultures, everywhere and at all times, have had names (Hough, 2016). We are now adding to those the names of new kinds of place, and with these new kinds of place come new kinds of names, as will be explored in chapter 6. The point here is that the newness of the names does not diminish the fact that they have their foundation in the human body, and that they are, alongside practices, ultimately indigenous to (in the sense of originating in) the body, and the body itself extends cognition outwards into the world in part through use of tools that include names, or implements for producing the permanence (e.g. making them stick through the use of a rod or map). It is to an exploration of some neurological and external structures for the generation and use of names and memes as tools (Wittgenstein, 2009) that we now turn.

Neurogeography

On April 17, 2013 Peter Moonias, Chief of the Neskantaga First Nation in Canada, declared a state of emergency (CBC, 2013). In the past several months, this northern indigenous community had experienced an unusual spike in the number of individuals taking their own lives. When asked by the CBC interviewer, "What do you think is to blame," an audibly upset Moonias responded, "there are many reasons... social issues from the past ... not only this community ... housing problems ... funding problems." Moonias went on to mention Canada's legacy of residential schooling, the fallout of which is still being felt on First Nations reserves and in communities across Canada in the form of drug and alcohol addiction and poor parenting skills. Broken continuities abound and intergenerational links, severed by the residential schools' active program (only ended in the 1970s in most places in Canada), are still being repaired (Niezen, 2013).

While many commentators, reporters, and academics from around the world have noted the phenomenon of suicide clusters in Canadian First Nations communities (Eades, 2014a; BBC, 2014; CBC, 2013; Evans, 2010; Chandler and Lalonde, 2009), others have indirectly corroborated a correlation of indigeneity and suicide in certain very specific places as exacerbated by the severing of memetic structures transmitted through time from parents to offspring (Leigh, 2010). The purpose of this chapter is first to describe how memetics, defined as the science of mapping intergenerational spatial and place-based knowledge systems (Eades, 2015), can be applied in relation to indigenous communities experiencing ongoing traumatic consequences of colonization and governmentalization of their lands, livelihoods, and ways of being.

Kirmayer et al. (2009, page 295) describe how, though consciousness is seated in the brain, indigenous health as tied to healthy minds extends well beyond the brain into bodies in interaction with places and spaces in environments that include traditional indigenous hunting grounds. As Stairs and Wenzel (1992) have also noted, the indigenous 'I' is much more: that 'I' (the singular conscious self) is 'I and the environment' that holds so much of what it means to be indigenous in Canada and around the world (Escobar, 2008).

The application of memetics to geographical and indigenous thought and life is fraught with risk. The meme is associated with the early reductionistic thinking of Dawkins (1976 and 1984), but later thinking on the subject has moved the meme beyond an association with genetics and biology, towards realms of extended cognition and cultural evolution (Distin, 2005 and 2010). At the same time, one of the meme's greatest champions is Dennett (1996), a neurophilosopher most famous for his computational theory of mind. This chapter places itself firmly within a non-reductionistic framework of extended-cognition thinking that does not eschew recent findings in neurology. The latter are considered too important for geographers to pass over (for recent work by geographers engaging with neuroscience see, for example, Gagen, 2013; Butcher, 2012; Laurier and Brown, 2008; Korf, 2008; McCormack, 2007).

Understandably, geographers have been critical of the uses of neuroscience as productive of a *scientistic* (as opposed to *scientific*, cf. Tallis, 2011) and governmental rationality that reduces learning to something associated with narrowly defined (and neurologically based) forms of happiness, contentment, and adjustment (Gagen, 2013). However, geographers' (Gagen, 2013; Butcher, 2012; McCormack, 2007) engagement with neuroscience itself, as a driver of both governmentalities and new forms of biopolitics, has been thin, relying upon two or three key figures from popular science, especially Damasio (1994, 1999, and 2003) and Ledoux (1996 and 2002), or upon post-structural thinkers using molecular/biological metaphors (McCormack, 2007), especially from Deleuze and Guattari (1987).

Geography's impoverished engagement with neuroscience is damaging because, on the one hand, neuroscience needs geography in order to escape the trap of reductionism, especially as critiques from within the discipline push neuroscientists to begin thinking about extended cognition (Tallis, 2011; Clark, 2011; Rowlands, 2010; Bergson, 1988). I call this kind of neuroscience with geography *neurogeography*. On the other hand, geography needs neuroscience most in the areas of therapeutics, affect, and mapping, as indicated by this chapter's opening observations of connections between identity, mental health, and land-based livelihoods (Kirmayer et al., 2009; Stairs and Wenzel, 1992).

In addition to neurological associations, memes implicate themselves as part of a paradigm of evolutionary thinking, again, one with which geographers have been hesitant to seriously engage. While papers on intergenerational learning are starting to appear in the geographical literature, they do not, for the most part, explicitly recognize their evolutionary provenance and heritage (van Ham et al., 2014; Hopkins et al., 2010). Exceptions to the anti-evolutionary trend in human geography exist (see for example Stallins, 2011; Eades, 2012a and 2015) but for the most part the ill-informed trend continues well after the 150th anniversary of the publication of Darwin's (1859) *Origin of Species* (Castree, 2009). As with neurology, evolutionary thinking has many potential applications and should be seen for what it is: a set of tools for thinking through geographical problems such as, for our purposes here, loss of indigenous identity and suicide, its prediction (Connor, 2014), and prevention (Leigh, 2010).

Place cells and the meaning of place

> [S]o where are you, you're in some hotel room … you just wake up and you're in a motel room. There's the key. It feels like maybe this is just the first time you've been there, but perhaps you've been there for a week, three months, it's kind of hard to say … it's just an anonymous room.
>
> From the film *Memento* (character of Leonard Shelby), Nolan 2001

For the purpose of beginning to map out areas of productive overlap between neuroscience and geography, the best place to start is the hippocampus. The rationale for starting here are 1) the hippocampus is the seat of the brain's cognitive map (O'Keefe and Nadel, 1978; Redish, 1997; Gluck and Myers, 2001), and 2) it provides an easily understandable introduction to the neuron, which is the basic unit of spatial information processing in the brain (Tallis, 2011; Zeman, 2008). Because of the size and shape of the neural structure of the hippocampus, neurons are seen most easily in this particular brain structure and are thus most easily mapped by neuroscientists. The hippocampus is also one of the only parts of the adult brain capable of generating new neuronal growth and, as such, it is uniquely situated as an area of focus for measured recovery from degenerative disease and stress (Anderson et al., 2006). The hippocampus is uniquely responsive to stress (or lack thereof), with potential for both long-term potentiation (LTP)—or the formation of new memories—and long-term depression (LTD) of hippocampal neurons in response to stress (Eichenbaum, 2002, page 63).

The much-touted cognitive map contained in the hippocampus is made up of place cells, each of which is uniquely responsive to location in space (O'Keefe and Nadel, 1978; Gluck and Myers, 2001). Oversimplifying for the purposes of building a picture of space from the ground up, we can say that the sum of neuronal firings that occur as a person moves around in space constitute a cognitive map of that space. For example, working with rats, scientists have been able to infer location from the firing of particular neurons in the rat hippocampus. As the hippocampus does not differ fundamentally between species, it can be further inferred that the same applies to humans. However, the use of neurons to map space is fraught with problems, first and foremost the problem of reducing meaningful place to sets of coordinates (or to neurons associated with those coordinates). There is admittedly the further problem of inferring human minds from those of animals, but it is beyond the scope of this book to do justice to this problem.

It is therefore incumbent upon the prospective neurogeographer to spell out how it is that the brain processes information about place while at the same time attempting to avoid some of the worst traps associated with both reductionism and the evacuation of meaning from established and hard-won ideas about place in the discipline of geography. The contemporary literature on place in geography begins with Tuan (1976/2001). Tuan differentiates between places imbued with meaning, memory, and affect and spaces that are more disembodied and quantitative representations seen as though 'from above' (i.e. as though in a map view). Cresswell is a more recent champion of the idea of place as meaningful and emotion-laden, with poetic import, in contrast to less meaningful spatial constructs that flow naturally from, for example, cartographic reason or ways of seeing associated with GIS (Cresswell, 2004 and 2013; Curry, 1998). In anthropology, Thornton has explored being and place as phenomenologically rich, and grounded local, human constructions with universal implications for cultural identity and territoriality (Thornton, 2008 and 2012).

In fact there is a smooth transition between the very small scale of the neuron and full-blown consciousness or mind and territorial identity (Zeman, 2008; Tallis, 2011) situated in place. It is possible, for example, to give a very clear and meaningful definition of place from the perspective of the hippocampus that is not incompatible with human geographical definitions of place. Gluck and Myers (2001, 30) have thus posited the following:

> What is a place? One definition is that a place is a collection (or configuration) of views. When we stand in one spot and look north and stand in the same spot and look south, those two views should be integrated into a unified percept of the current location so that the next time we approach that spot (from any angle), we recognize where we are. In addition to visual cues, there may be auditory, olfactory and tactile cues, as well as memory of the route by which we reached the spot and what happened when we got there. All this information should be combined into the memory of a 'place.'

As indicated by Leonard Shelby ('Lenny') in the movie *Memento*, without the hippocampus, human beings have no way of synthesizing perspectives about places because they have no memory of those perspectives. It is precisely this kind of episodic memory (as opposed to procedural memory of how to perform mechanical and non-spatial tasks) that the hippocampus enables and it is therefore of central importance to the project of neurogeography. Without the hippocampus, episodic memory, and thus memory of places, is absent. Exploration of the hippocampus is therefore necessary, but not sufficient, for building a stable platform upon which neurogeography will be built. What is required beyond recognition of the brain's place for places is how the hippocampus coordinates neural activities towards establishing meaning across multiple places in sequence, and therefore towards meaningful embodied activity across both time and space. This, in turn, has implications for indigenous senses of health and recovery through land-based activity as a way of overcoming damaging colonial and governmental policy legacies.

A memetics of place

With the brain's 'place for places' established in the hippocampus, we move now to an exploration of how those brain places translate into memes which are 1) material structures in the brain beyond the single neuron, i.e. neural groups or cliques selected through repeated use, and 2) performances involving bodies in interaction with the world and its (meaningful) places.

The material structure of a meme in the brain is formed through repetitive firing of groups of neurons characterized by Edelman (1987) as neural groups; and by Leigh (2010) as neural cliques. The latter is most explicit about such neural clusterings providing a material basis for memes defined as:

functional neuronal clusters (*neural cliques*) with long term potentiation. Meme replication occurs in the brain when a meme containing neuronal cluster is reinforced by stimulation, which may in turn infect other clusters to change configuration.

(Leigh 2010, 21, italics in the original)

While Leigh's definition is useful in defining the structure of memes in the brain it is, however, Edelman who gives the most clarity about the evolutionary logic behind the selection of neural groups. According to Edelman (1987 and 1994) neurons are selected in mutually reinforcing clusters through group activation during a variety of brain activities corresponding to perceptions and intentions of the bodies in which they are housed. Through repetitive activation of these neural groups (corresponding to repetitive activities and perceptions) those groups are selected as units through natural selection (Godfrey-Smith, 2009). However, as we will see below, it is *cultural* selection that provides a constraint upon natural selection, and which leads away from an environmentally determinist basis for the meme. Instead, the meme is culturally constrained in the brain of its host that must, all the same, operate according to physical laws at the neural level.

For the purposes of this chapter, the Leigh/Edelman model of neural clustering and selection is used as the basis for material structures for memes in the brain, towards establishing how memes, defined as units of cultural information, are performed on the ground. Interaction with multiple places in sequence, or a journey, fires sets of neural clusters in specifically patterned ways that activate long-term potentiation, or meaningful memory formation. For example, one might return to the same place each year for holidays. The path taken towards that place might stay the same over several years, based upon considerations of traffic congestion, scenery, and stops for petrol or food. The point here is that some process of selection guides the choice of route that remains stable over time.

Variations can strengthen route choice as better routes are found, or as, for example, road construction occurs or new stopping places are built. This abstract example of holiday routes has somewhat limited applicability towards indigenous peoples and mental health, despite indigenous tourism, for example, being an important component of the theory being put forward here (Eades, 2015). We therefore explore below a real-world example one that goes beyond simple notions of tourism as dominated by outsiders, from the Cree community of eastern James Bay, that of a yearly commemorative event in Wemindji, Quebec known as *kaachewaapechuu*. This event is posited as having both spiritual and therapeutic value for its participants across a range of ages from youth to adults and elders. *Kaachewaapechuu* is a culturally selected route that is performed (with small variations depending on changing conditions) in the same way each year. Multiple generations of Cree and non-Cree participants complete the walk each year. It is an episode in the life of the community that is seen as extremely positive in terms of both health and the formation of new positive memories.

Kaachewaapechuu and extended cognition

> A fissure emerges between cognitive mapping and orientation.
>
> (Manning 2009, 168)

Kaachewaapechuu means, literally, 'going offshore', but I came to think of it, during my participation in the event in 2010, as 'the long walk'. It is part of Cultural Awareness Week held in Wemindji each year. Cultural Awareness Week is itself part of a larger set of structures and institutions in Wemindji that have given this community a reputation for being healthy. In addition to programs such as this, held in the local community centre, Wemindji has a school, fire and police services (with dedicated hotlines for emergencies), and a government and services office staffed with a balanced representation of both male and female Cree individuals with specializations in social issues, mapping, and economic development, to name just a few. It is widely known in the eastern James Bay area that Wemindji is the last 'untouched' community (meaning that it is relatively unaffected by large scale resource development schemes) and that it therefore has a sense of intimacy lacking from larger centres in the area (e.g. Chisasibi) but even this is changing in recent years as mining companies, for example, begin to zoom in and bring a few jobs with them (Scott, 2001). Figure 4.1 shows supportive community members who turned out to help us begin our journey by offering morale-boosting advice.

Kaachewaapechuu involves a three-day walk along the James Bay coast from Wemindji to an old village site that was abandoned for the newer,

Figure 4.1 The start of *kaachewaapechuu* at Wemindji

contemporary Wemindji, site fifty years ago (Cree Nation of Wemindji, 2010). The idea of a 'starting point' is, for the purposes of this chapter, somewhat problematic. We began by discussing a neurological basis for memes but this is not intended as a statement of ontological priority of brain-based explanations. Instead, we posit modern-day Wemindji as the platform from which community health begins, exemplified symbolically by *kaachewaapechuu* and Cultural Awareness Week (of which *kaachewaapechuu* is a part), with knock-on effects for both individual and community senses of health, self, and well-being. Health and prosperity are, after all, the intended outcomes of local cultural and educational institutions in Wemindji. For the purposes of neuro-geographical explanation, it is not sufficient to know that these institutions work. The neurogeographer will ask why they work and, therefore, we begin to build a holistic framework for providing this necessary explanation, one that builds upon a neurologically grounded memetic theory (as outlined above), toponymy, and extended cognition (Clark 2011). These are part and parcel, in turn, of building up a detailed picture of exactly how, and the mechanisms by which, the tools and NTNs that are the subject of this book work.

The long walk from Wemindji to the old village site abandoned, for the most part, fifty years ago begins with maps and place names. When I took part in this event we sat with an elder whose trapline and cabin were situated at the end of the first day's walk. This elder spread a map before us on a table in the community centre in Wemindji and proceeded to explain to us in Cree (we had help from a translator) how the journey would proceed. The path was traced many times by his finger, noting constraints in the paths along which our sleds would pass. We would cross over frozen lakes, ponds, and bogs and the going would therefore be much easier this winter season than it would have been had the journey taken place in summer (Figure 4.2). At the end of the first day we would bunk down, all eleven of us (including a young Cree man and his two children, a half-Cree woman and her child, three non-Cree residents of Wemindji, two Cree adult women, and an elder Cree woman), in this trapper's cabin (Figure 4.3). It was his job to check on us throughout the journey each of the three days of the walk to ensure our safety and progress, passing by us every couple of hours on his snowmobile.

As the participants' gaze settled upon the map, that map, as a representation of the landscape, became involved in our thinking and the elder's finger tracing trajectories through features named in the Cree language, our thinking looped through that finger, the language, mixing with bodily anticipation of the journey ahead. Below, as the path of *kaachewaapechuu* for 2010 is described, the argument is made that thinking is not just in the brain but also performed both bodily and in interaction with landscape and its places, spaces, and representations shown on the map. This is a jump from the neurological-level explanation of things, alluded to by the Manning quote that opened this section. It behoves us to maintain awareness of just how many levels of consciousness are involved in planning and executing a 40-km walk in Eastern James Bay, Quebec, Canada, spanning multiple places, and crossing icy bays,

Figure 4.2 Crossing a frozen bay

Figure 4.3 Interior of trapper's cabin

shorelines, and wide expanses of snow and rock. Every rock, shoreline, and bay is, for the Cree hunter, a part of thinking in and through the land and its inhabitants: humans, caribou, fox, rabbit, and ptarmigan, to name but a few (Berkes, 2008; Carlson, 2008; Morantz, 2002; Scott, 1983). This involves not only neurological and bodily coordination but an actively intentional stance amongst the participants that very much goal-directed but not deterministically so (Tallis, 2011; Dennett, 1989).

I had brought my GPS unit for this walk and it further guided the thinking of our group, acting at times almost like a replacement spatial indexer for our hippocampi (or at least my non-Cree hippocampus). But the GPS was one among many devices (including paper maps) for planning and structuring our walk. It often elicited the question "How far?" and I would read the straight-line distance to our destination for the day. The GPS was also loaded with a database of 898 Cree place names. When approaching a place, it would become visible as a point upon the small screen, and I could read the Cree name out to someone for translation. For example, at the end of our second day walking I read the name *aamuutaayiminaanuwich* (strawberry-eating place) to the young Cree man (and future trapline boss) traveling with his two children. He provided the translation for this waypoint, not far from his uncle's cabin, a place where at least one Cree family maintains year-round (though sporadic) life on the land. The GPS provides an example of one aspect of 'extended mind', of a device for 'thinking through' landscape and environment (Clark 2011; Rowlands 2010).

Waypoints, maps, movement, and group discussions alongside devices like the GPS, or brain-based methods for 'knowing' location (such as activation of hippocampal place-sensitive neurons) all form part of the intentional stance of participants in *kaachewaapechuu*. There are a great number of 'levels' involved in making one's way across forty kilometers of Canadian Shield snow, forest, and bay including at least the following sub-systems: neurological, memory, body, logico-deductive, inductive, inscriptive (i.e. maps and GPS), and environmental. For the purposes of this paper I posit all of the above (and surely there are more) to be involved in some kind of formulation of extended mind. Zeman (2008), Chalmers (2010), and Clark (2011) have discussed at length how various levels 'link up' across a range of scales from the molecular to the extra-bodily and beyond. We posit here the named place, i.e. each with at least one place name or recorded reference, as key components of extended, mind-based, consciousness of the environment. Furthermore, it is one that is linked directly in, at a most fundamental level, to a neurological basis for memory. In other words, the place name corresponds at the level of the hippocampus to the place cell, seat of episodic (i.e. of event-based) memory in the brain. These place cells are like spatial indexers providing the guiding correlation between brain, body, and land so necessary for successful navigation and wayfinding in complex environments (Godfrey-Smith, 1996).

I have already described the long walk in detail—especially in terms of cultural practices and intergenerational knowledge transmission—in some of my previous work (Eades, 2012b and 2015). The purpose here is not to go over already-covered academic territory but to extend that territory into previously unexplored realms of extended cognition and neurological geographies, or neurogeographies. *Kaachewaapechuu* remains an exemplary case study for doing so. But surely neurogeography and the mapping of *kaachewaapechuu* is not merely about mapping correlates between interactions with places in the landscape and correlated neural firings in the brain. The beauty of the ongoing revolution in cognitive *science* is that it is beginning to map out ways in which mind exists not only in the brain but also in the body, in artefacts, and in the environment itself (Rowlands, 2010). Tim Ingold has made this point consistently in writing about hunter-gatherer cultures, including, notably, the Cree (Ingold, 1986, 2000, 2007, and 2011).

The new science of mind (Rowlands, 2010) posits mental processes that overlap, or spill over, to varying degrees into the environment, objects, and inscriptions. An example of a mental process that spills over into the environment and is inscribed is a map. Rowlands (2010, pages 13–19) uses the examples of GPS and MapQuest to illustrate the point, examples that are especially illuminating for the discussion of *kaachewaapechuu* as a mapped phenomenon performed by bodies (i.e. the eleven participants) in part with the help of maps and, because I brought the unit with me, GPS. GPS enables a mind to offload information into an external device, thus reducing the cognitive load for the traveler. In Rowlands words, "the number and complexity of neural operations that I must perform to accomplish the task of getting from A to B is accordingly reduced" through the use of a map (and by extension by the use of GPS). For Cree travelers, GPS is not necessary because at least one of the group has always performed the meme (i.e. the set of linked and replicable sets of named places traveled to in sequence) before. For non-Cree or outsiders like myself, GPS provides a database that supplements and offloads my own cognitive structures and memories, including some of those found in the hippocampus.

Rowlands (2010, page 18) makes the further point that externally inscribed maps on paper or, indeed, in GPS, transform information that is merely present to that which is available, in and through the manipulation of maps. What this means is that after a person has been trained to use maps by, for example, learning how to re-orient the map based on the north arrow (and corresponding environmental knowledge of where north actually is in the current situation), the map goes from being a *mere* inscription to something with informational value. When I encountered the place name *aamuutaayiminaanuwich* on my GPS unit, I learned how to decode that information through another mind, that of a Cree co-participant and friend, the future trapline boss who was accompanying his two children to *maatuskaach siipii* (lots of poplars river), where the group would be stopping at the end of the second day of the walk.

The third day of *kaachewaapechuu* turned out to be one of the hardest because of an accumulation of trials the day before. That day we had crossed a very long, indeed seemingly endless, expanse of frozen bay: the far shore appeared to recede with every step (see Figure 4.2). The children in the group often stopped to check in with parents, play, and then run off ahead again.

The children's spirit of playfulness seeped into every aspect of our journey, but because of the extra energy expended through play, many of the younger members of our traveling group were quite tired by the third day. This tiredness also seemed to seep through the group. In contrast, the oldest member, a female elder who had completed *kaachewaapechuu* many times before, embodied the saying "Slow and steady wins the race." Though each of her footsteps seemed to cover very little ground, the frequency of those steps and the fact that she never lost ground or took breaks meant that she was often way out ahead of everyone else.

We nevertheless managed to catch up with her now and then to have a chat about our progress, discuss the timing of the next meal, or talk about maps. This eldest member of our group began to habitually ask me "How far?"—partly as a joke aimed at one of the slower members of the group (me, though to be fair, compared to her we all fell into that category) but also as an excuse to begin talking about our surroundings. The GPS became a sort of evocative object (Turkle, 2007) providing seeds for other conversations, thoughts, and observations about trees species, food, snow depth, and travel speed, which were all related in her (and increasingly in my) mind. But more importantly, this elder served as an example and a model of locally situated indigenous knowledge. Even during play, children would pause to notice or listen to the conversation of adults/elders, to observe or imitate their gestures or styles of speech. This occurred constantly not only during the long walk but also afterwards when we had settled down in the cabin at Old Factory Bay, where some of the very young children had access to a satellite TV beaming in cartoons such as *SpongeBob SquarePants*, which seemed almost surreal at the end of such a journey.

Even while watching something as silly as a cartoon, children listened to the talk and activities around them—for instance, to an older man telling me about how some of his ancestors had died of starvation in a cabin not far from where we sat eating stew and drinking hot tea. The TV and the elder's tragic story competed for the young children's attention (some as young as three or four, many of whom had not accompanied us on our walk but were there to greet us at the end of it), literally pulling their bodies this way and that across the floor of the large cabin where at least ten people slept at night. During the day this was a place of hustle and bustle, at times noisy and clamorous and at other times with outside sounds filtering in—such as the daily plane from Wemindji passing over the cabin towards Eastmain, Val d'Or, and, ultimately, Montreal. The children were aware of all of this and were much more at ease in their surroundings than was I during those first couple of hours. It all seemed to me, at least at first, to be very chaotic.

The situation at the end of the long walk was not, however, all that different from that which pertained during it. As we performed *kaachewaapechuu* by moving from place to place, the children had been in a constant state of play, occasionally pulled into the older peoples' group through want of attention or a need for food, water, or rest (Figure 4.4). The main point is

Figure 4.4 Youth resting with sled dog

that there was a great deal of listening going on, and it was a two-way street. In Wemindji, children are taught to obey their elders. At the same time, as Cultural Awareness Week exemplifies, there is a great deal of listening the other way, of elders listening to the needs of youth. On our return to Wemindji, a couple of days after the long walk, I produced a map of *kaachewaapechuu* by draping the GPS track across Google Earth imagery. This map was printed on four letter-sized (A4) sheets placed end to end and entered in the yearly art contest that is part of Cultural Awareness week in Wemindji (Figure 4.5). This art contest is a way for contestants in four categories (children, youth, adults, and elders) to express ideas about landscape in a visual form for others to contemplate and, ultimately, judge (Eades 2015).

Memetic misgivings?

The problem with the above, one might posit, is that it could be seen to offer a confused sense of the 'unit' of analysis being used to conceptualize place-based practices. This unit could be seen, variously, as a select set of neurons in the brain, a linked set of place names that correlate to those selected neurons, the named places (referents) themselves, the entire journey of *kaachewaapechuu*, its performance, its inscription on maps, the GPS, the individual body in motion, or the complete set of participants (and their performances) undertaking the long walk from Wemindji to Old Factory Bay. The problem of defining the unit of analysis is inherent not only to memetics (see Aunger,

Figure 4.5 Wemindji art contest with *kaachewaapechuu* Google Earth map

2002 for a good overview), but also to the closely related field of assemblage (see Eades and Zheng, 2014; DeLanda, 1997, 2002 and 2006; Deleuze and Guattari, 1994), and is usually dealt with by focusing on interchangeability of representational forms (Distin, 2005). The problem is summarized aptly by Ingold (2000, 163) in terms of a critique of the cognitive science paradigm applied to anthropology, in which,

> [w]hereas cognitive scientists ... have by and large been concerned to discover universals of human cognition, which are attributed to innate structures established in the course of evolution under natural selection, cognitive anthropologists have sought to account for human perception and action in terms of acquired schemata or programmes that differ from one culture to another.

The problem with such a view (according to Ingold), that of seeing minds as universal computer programs played out on the 'hardware' of the brain, is that it is both deterministic and reductionistic of human potentiality. In essence, the cognitive scientist has moved past behaviorism with its black boxes; now the boxes are filled with programs and named according to functional criteria. The hardware in question would be the hippocampus, and the software a cognitive map that is generated through interaction with sets of named places (e.g. Wemindji, *aamuutaayiminaanuwich, maatuskaach siipii,* Old Factory Bay). Each area of the brain, considered separately, corresponds to one and only one kind of function, running specific kinds of program. According to this logic the hippocampus is responsible for running one kind of 'program' for generating cognitive maps. The limitations of this way of seeing things are obvious. For example, by what means are the various functions and programs bound together to give the brain, mind, and subject their characteristic capabilities to plan for the future, foresee problems and overcome them—in other words, to act intentionally (Tallis, 2011, 134)?

It is beyond the scope of this chapter (and book) to address this question, and cognitive scientists themselves are still far from having an answer to the 'binding' problem of mind and consciousness, one that brings together ('binds') all the various 'programs' with overlapping and multiple functionings into a cohesive and coherent whole. It does, however, point to a related problem, that of seeing cultures, in the absence of universal structures and schema, as exhibiting practical and behavioral characteristics relative to a continuum of practices and behaviors varying across space and time amongst various other peoples and cultures. This 'opting out' version of anthropology throws out the universal bathwater with the baby of cognitive science, despite recent advances in the latter that are beginning to move beyond the computer analogy (Malafouris, 2013; Clark, 2011; Menary, 2010; Rowlands, 2010; Manning, 2009; Kandel, 2006; Noe, 2004). These advances are beginning to bridge areas of philosophy and geography formerly considered beyond the reach of cognitive science, with its implications of computational representationalism. There is a growing realization that computation and representation, while not the panaceas they once were considered to be, still hold a place in developing phenomenologies of place, space, and being. Non-representational geographies (Thrift, 2007), for example, could learn much from paying attention to cognitive science as it matures into realizing more nuanced versions of itself that give representation and computation limited or, better, very specifically defined roles in theories of extended mind, phenomenology, and non-reductionistic theories of being and place (Rowlands, 2006 and 2010; Malpas, 2008).

In other words, it does not make sense to idealize by relativizing indigenous communities posited as 'each doing their own thing'. In Canada, communities experiencing suicide clusters have structural features in common, and these, in turn, are related to historical factors, colonial and governmental practices and, often, the failure to create positive institutional factors for countering negative outcomes of such practices. Residential schooling, community displacement, and the removal of children from their families in northern and indigenous communities were ended only in the latter half of the twentieth century. The experience of severed ties to the land, language, and family was bundled together in a program of assimilation the effects of which are still felt today across Canada. This negative legacy, alongside efforts to correct the damage, are universal experiences and need to tie into universal ways of seeing mental health, from the reductionistic (brain-based) to the holistic (mindfulness on the land). The presence of positive institutions of eight distinct types (self-government, land claims, education, health, cultural facilities, police and fire, women in government, child services) (Chandler and Lalonde 2009, 240), combined with identity-, land-, and brain-based ways of conceptualizing health that are in turn tied to physical activity for the purposes of gaining sustenance, mental stimulation (as opposed to boredom induced by sedentary town life), and renewed senses of the value of indigenous conceptions of the cosmos, all lend themselves to seeing the inherent value of an activity such as *kaachewaapechuu*. The latter is symbolic of a problem that seems to affect especially unemployed men in the north. In speaking of brain-based disorders among Inuit of Quebec and lack of raw bloody meat in the modern diet, Kirmayer et al. (2009, 295) note, "[t]his is particularly a problem among men for whom being outside and moving across large territories while hunting has been an important element of experience and identity buttressed by cultural, linguistic, and cosmological principles."

I posit here that such an experience is near to being universal among both men and women (who contemporarily hunt and fish alongside or in addition to men) in northern indigenous and hunter-gatherer communities. A significant portion of the problem can be understood by looking not only at how brains function in different settings but also at how the level of physical activity that houses the brain represents itself to itself and others and thereby creates either positive or negative memes (Leigh, 2010). This, as I have demonstrated above, is intricately keyed in to places, place names, and activity on the land in conjunction with cultural awareness and positive institutions. There are implications of this way of theorizing (and naming) place across national boundaries as well (i.e. beyond Canada) to anywhere that colonizing legacies have left their marks on the maps (and brains) of local cultures.

Memetic awareness and neurological health

We pass, by imperceptible stages, from recollections strung out along the course of time to the movements which indicate their nascent or possible action in space. Lesions of the brain may affect these movements, but not their recollections.

(Bergson 1988, page 79; italics in the original)

Given the objections posed above, the incipient neurogeographer must proceed carefully, avoiding the trap of computationalism, a form of determinist reductionism, on the one hand, and that of idealizing the indigenous 'other,' a kind of pan-indigeneity, on the other. Wemindji is in many ways a model community, but it is not without the social problems that afflict many other northern communities in Canada. Police in Wemindji report that drug and alcohol abuse—in turn fallout from dysfunction due to isolation, boredom, and broken continuities in family life—often lead to violent crime. And knowledge of how the brain works will be useless in the face of the suicide of young people that occurs with higher frequency in other communities but is not unknown in Wemindji. What a memetic view of the brain and body will help bolster is a holistic view of indigenous life that is adapting and evolving in the face of rapid cultural change (Scott, 2001; Berkes, 2008).

One of the most important implications of a theory of extended mind based in neurology that undergirds a memetics of place is that cognition is in part constituted by aspects of the environment, including not only named places, topographies, and aspects but also *other minds*. Those other minds exist in various states that have an effect on those around them, and in rapidly changing and centralizing indigenous communities across the north (Eades 2013), proximity of other minds is of paramount importance, and the health of bodies, brains, and actions that constitute them of equal weight. Individual and community sense of health hinges upon this realization, and without such a sense of proximal and extended aspects of health, suicide clusters such as experienced in Neskatanga, and many other communities in Canada over several decades will remain a reality. This is an uncontroversial assertion, but one that is made all too infrequently in the geographical literature. Detailed and ethnographically informed examples of positive contributions to indigenous health and institutions are needed, and one need look no farther than the many healthy communities, of which Wemindji is but one example, for a variety of activities, methods, and initiatives for challenging negative legacies and stereotypes not only of Canada's indigenous peoples but of resonant struggles around the world (Niezen, 2010; Chatwin, 1987).

As neuroscience and philosophies of extended and embodied cognition and health develop, geographers can grow in tandem or in sympathy with these developments. If we fail to reap the rewards of these growing fields of inquiry and knowledge, we fail those who depend upon us as well—educators, academics, and knowledge practitioners including indigenous peoples and ourselves alike (Zeman, 2008). Without looking to these fields, geography will remain intractably internalist, with publication in the field an endless self-referential regression of non-accumulation (Eades, 2012a). Geographers need not fear the accumulation of positive knowledge; foundational discoveries in neurological inquiry remain just that: foundations or platforms from which better informed and more effective decisions about geographically related or influenced disorders can be made. These include those located in bodies, in the environments in which they operate, and in the minds that they both embody and integrate into worlds.

Nor does this chapter represent an attempt to precisely locate or provide spatial boundaries around cognition. Rowlands (2010, page 83) has warned that we must instead look to the external world for clues as to how minds work. In doing so, critical geographers have less to worry about with regard to perceived positivism of neuroscientific inquiry. Critical inquiry in political, human, and cultural geography has in large part been concerned with projects that take issue with absolute or reductionist theories of place, location, mobility, and boundedness (Adey, 2009; Pickles, 2004; Cresswell, 2006; Dalby, 2002). The theoretical positioning of neurogeography and a memetics of place seeks to extend these valuable projects, taking from neurophilosophical debates insights applicable to, and in sympathy with (rather than in opposition to) the spirit of critical inquiry in geography. This research also extends into, and begins to fill, a notable gap, namely, a shortage of studies looking at indigenous and non-western cultures, in critical geographical inquiry (Blomley, 2014; Sletto, 2014).

Though the title of this chapter indicates neurogeography as the primary concern of the research, the use of this new word is intended as a two-edged sword, as it were, with both reductive (neurological) and holistic (geographical) connotations. At the same time, a memetics of place embraces both levels at once and includes performed (in terms of bodily movement across the land), inscribed (maps and texts), and intergenerational (communicated and trans-mitted) aspects, exemplified in the yearly long walk (*kaachewaapechuu*) in Wemindji, Quebec. These in turn are parts of geo-NTNs, sets of tools for tracking place-based change. This community in many ways represents a picture of good health due to events like the long walk which exist in large part, and have symbolic value, because of the existence of healthy institutions (Chandler and Lalonde, 2009, page 240).

Memes are underutilized in geography (Eades, 2011 and 2015) for reasons noted above, but this situation need not persist, especially given the potentially productive nature of a memetics of place for human geographers, cognitive scientists, health and educational professionals, and indigenous peoples. Because place-based memes are theorized here as holistic devices for literally performing good health they are also applicable to humanity as a whole, for whom the idea of a memetics of place is already somewhat natural in tourism, commemorative travel, and outdoor activity (though not usually explicitly formulated as such). Meaningful place-based activities revolve around stable named features in the landscape (whether a city or a natural setting), inter-generational participation, and repetitive performance and these, it has been demonstrated above, are near-universal features of human spatial activity (Groh, 2014). For indigenous peoples specifically and especially in acute situations such as those experienced by the community of Neskatanga mentioned at the start of this chapter, such formulations can start to (per)form part of community healing programs and institution building.

References

Adey, Peter. 2009. *Mobility*. London and New York: Routledge.

Anderson, Per, Morris, Richard, Amaral, David, Bliss, Tim and O'Keefe, John (eds). 2006. *The Hippocampus Book*. Oxford: Oxford University Press.

Auge, Marc. 2009. *Non-Places: An Introduction to Supermodernity*. London: Verso.

Aunger, Robert. 2002. *The Electric Meme: A New Theory of How We Think*. New York: The Free Press.

BBC 1. 2014. A Cabbie Abroad. www.bbc.co.uk/programmes/b048s509 (accessed July 30, 2014).

Bergson, Henri. 1988. *Matter and Memory*. New York: Zone Books.

Berkes, Fikret. 2008. *Sacred Ecology*. London and New York: Routledge.

Blomley, Nicholas. 2014. The Ties That Bind: Making Fee Simple in the British Columbia Treaty Process. *Transactions of the Institute of British Geographers*. 40(2). 168–179.

Butcher, Stephen. 2012. Embodied Cognitive Geographies. *Progress in Human Geography*. 36(1). 90–110.

Carlson, Hans. 2008. *Home is the Hunter: The James Bay Cree and Their Land*. Vancouver: UBC Press.

Castree, Noel. 2009. Darwin and the Geographers. *Environment and Planning A*. 41(10). 2293–2298.

CBC online. 2013. Suicide Crisis in Northern Ontario. www.cbc.ca/player/Embed ded-Only/News/ID/2380433781/ (accessed August 1, 2014).

Chalmers, David. 2010. *The Character of Consciousness*. Oxford: Oxford University Press.

Chandler, Michael J. and Lalonde, Christopher E. 2009. Cultural Continuity as a Moderator of Suicide Risk among Canada's First Nations, in Kirmayer, Laurence and Valaskakis, Gail (eds). *Healing Traditions: The Mental Health of Aboriginal Peoples in Canada*. Vancouver: UBC Press.

Chatwin, Bruce. 1987. *The Songlines*. London: Vintage.

Clark, Andy. 2011. *Supersizing the Mind: Embodiment, Action, and Cognitive Extension*. Oxford: Oxford University Press.

Connor, Steve. 2014. Blood Test for Suicide Risk in Soldiers. *The Independent on Sunday*. 3 August 2014: 24.

Cowen, Rob. 2015. *Common Ground*. London: Hutchinson.

Cree Nation of Wemindji. 2010. *Wemindji Turns 50*. Milton: Farrington Media.

Cresswell, Tim. 2004. *Place: A Short Introduction*. Malden: Blackwell.

Cresswell, Tim. 2006. *On the Move: Mobility in the Modern Western World*. London and New York: Routledge.

Cresswell, Tim. 2013. *Geographic Thought: A Critical Introduction*. Chichester: Wiley-Blackwell.

Curry, Michael. 1998. *Digital Places: Living With Geographic Information Technologies*. London and New York: Routledge.

Dalby, Simon. 2002. *Environmental Security*. Minneapolis: University of Minnesota Press.

Damasio, Antonia. 1994. *Descartes' Error*. New York: Putnam.

Damasio, Antonio. 1999. *The Feeling of What Happens: Body, Emotion, and the Making of Consciousness*. New York: Harcourt Brace.

Damasio, Antonio. 2003. *Looking for Spinoza: Joy, Sorrow, and the Feeling Brain*. London: William Heinmann.

Darwin, Charles. 2010. *Evolutionary Writings*. Oxford: Oxford University Press.

Dawkins, Richard. 1976. *The Selfish Gene*. Oxford: Oxford University Press.

Dawkins, Richard. 1984. *The Extended Phenotype*. Oxford: Oxford University Press.

DeLanda, Manuel. 1997. *A Thousand Years of Non-linear History*. New York: Zone Books.

DeLanda, Manuel. 2002. *Intensive Science and Virtual Philosophy*. London and New York: Continuum.

DeLanda, Manuel. 2006. *A New Philosophy of Society: Assemblage Theory and Social Complexity*. London and New York: Bloomsbury.

Deleuze, Gilles and Guattari, Felix. 1987. *A Thousand Plateaus: Capitalism and Schizophrenia*. Minneapolis: University of Minnesota Press.

Deleuze, Gilles and Guattari, Felix. 1994. *What is Philosophy?* London and New York: Verso.

Dennett, Daniel. 1989. *The Intentional Stance.* Cambridge, MA: MIT Press.

Dennett, Daniel. 1996. *Darwin's Dangerous Idea.* New York: Simon & Schuster.

Desbiens, Caroline. 2013. *Power from the North: Territory, Identity, and the Culture of Hydroelectricity in Quebec.* Vancouver: UBC Press.

Distin, Kate. 2005. *The Selfish Meme: A Critical Reassessment.* Cambridge: Cambridge University Press.

Distin, Kate. 2010. *Cultural Evolution.* Cambridge: Cambridge University Press.

Eades, Gwilym. 2011. Place Memes, presentation to the annual Association of American Geographers meeting, Seattle, Washington.

Eades, Gwilym. 2012a. Determining Environmental Determinism. *Progress in Human Geography.* 36(3). 423–427.

Eades, Gwilym. 2012b. Cree Ethnogeography. *Human Geography.* 5(3).

Eades, Gwilym. 2013. Toponymic Constraints in Wemindji. *The Canadian Geographer.* 58(2). 233–243.

Eades, Gwilym. 2014. Self-Curation and the Memetic Academic, presentation to the annual Royal Geographical Society meeting, London.

Eades, Gwilym. 2015. *Maps and Memes: Redrawing Culture, Place, and Identity in Indigenous Communities.* Montreal and Kingston: McGill-Queen's University Press.

Eades, Gwilym and Zheng, Yingqin. 2014. Counter-mapping as Assemblage: Reconfiguring Indigeneity, in Doolin, Bill, Lamprou, Eleni and Mitev, Natalie (eds). *Information Systems and Global Assemblages: (Re)configuring Actors, Artefacts, Organizations.* Heidelberg: Springer. 79–94.

Edelman, Gerald. 1987. *Neural Darwinism.* New York: Basic Books.

Edelman, Gerald. 1994. *Bright Air, Brilliant Fire.* New York: Basic Books.

Eichenbaum, Howard. 2002. *The Cognitive Neuroscience of Memory.* Oxford: Oxford University Press.

Escobar, Arturo. 2008. *Territories of Difference: Place, Movement, Life, Redes.* Durham: Duke University Press.

Evans, Al. 2010. *Chee Chee: A Study of Aboriginal Suicide.* Montreal and Kingston: McGill-Queen's University Press.

Gagen, Elizabeth A. 2013. Governing Emotions: Citizenship, Neuroscience and the Education of Youth. *Transactions of the Institute of British Geographers.* 40(1). 140–152.

Gluck, Mark A. and Myers, Catherine E. 2001. *Gateway to Memory: An Introduction to Neural Network Modeling of the Hippocampus and Learning.* Cambridge, MA: MIT Press.

Godfrey-Smith, Peter. 1996. *Complexity and the Function of Mind in Nature.* Cambridge: Cambridge University Press.

Godfrey-Smith, Peter. 2009. *Darwinian Populations and Natural Selection.* Oxford: Oxford University Press.

Groh, Jennifer M. 2014. *Making Space: How the Brain Knows Where Things Are.* Cambridge, MA: Harvard University Press.

Grossman, Stephen. 2011. Cortical and Subcortical Predictive Dynamics and Learning during Perception, Cognition, Emotion, and Action, in Bar, Moshe (ed.). *Predictions in the Brain: Using Our Past to Generate a Future.* Oxford: Oxford University Press.

Hanna, Patricia and Harrison, Bernard. 2004. *Word & World: Practice and the Foundation of Language.* Cambridge: Cambridge University Press.

Hopkins, Peter, Olson, Elizabeth, Pain, Rachel and Vincett, Giselle. 2010. Mapping Intergenerationalities: The Formation of Youthful Religiosities. *Transactions of the Institute of British Geographers.* 36(2). 314–327.

Hough, Carole (ed.). 2016. *The Oxford Handbook of Names and Naming.* Oxford: Oxford University Press.

Ingold, Tim. 1986. *Evolution in Social Life.* Cambridge: Cambridge University Press.

Ingold, Tim. 2000. *The Perception of the Environment.* London and New York: Routledge.

Ingold, Tim. 2007. *Lines: A Brief History.* London and New York: Routledge.

Ingold, Tim. 2011. *Being Alive: Essays on Movement, Knowledge and Description.* London and New York: Routledge.

Kandel, Eric. 2006. *In Search of Memory: The Emergence of a New Science of Mind.* New York and London: W.W. Norton.

Kirmayer, Laurence, Fletcher, Christopher and Watt, Robert. 2009. Locating the Eco-centric Self: Inuit Concepts of Mental Health and Illness, in Kirmayer, Laurenceand Valaskakis, Gail (eds). *Healing Traditions: The Mental Health of Aboriginal Peoples in Canada.* Vancouver: UBC Press.

Korf, Benedikt. 2008. A Neural Turn? On the Ontology of the Geographical Subject. *Environment and Planning A.* 40(3). 715–732.

Kripke, Saul. 1981. *Naming and Necessity.* Malden: Blackwell.

Laurier, Eric and Brown, Barry. 2008. Rotating Maps and Readers: Praxiological Aspects of Alignment and Orientation. *Transactions of the Institute of British Geographers.* 33(2). 201–216.

Ledoux, Joseph. 1996. *The Emotional Brain.* New York: Simon & Schuster.

Ledoux, Joseph. 2002. *Synaptic Self: How Our Brains Become Who We Are.* London: Penguin.

Leigh, Hoyle. 2010. *Genes, Memes, Culture, and Mental Illness.* New York: Springer.

Malafouris, Lambros. 2013. *How Things Shape the Mind: A Theory of Material Engagement.* Cambridge, MA: MIT Press.

Malpas, Jeff. 2008. *Heidegger's Topology: Being, Place, World.* Cambridge, MA: MIT Press.

Manning, Erin. 2009. *Relationscapes: Movement, Art, Philosophy.* Cambridge, MA: MIT Press.

McCormack, Derek P. 2007. Molecular Affects in Human Geographies. *Environment and Planning A.* 39(2). 359–377.

Menary, Richard (ed.). 2010. *The Extended Mind.* Cambridge, MA: MIT Press.

Morantz, Toby. 2002. *The White Man's Gonna Getcha.* Montreal and Kington: McGill-Queen's University Press.

Niezen, Ronald. 2010. *Public Justice and the Anthropology of Law.* Cambridge: Cambridge University Press.

Niezen, Ronald. 2013. *Truth & Indignation: Canada's Truth and Reconciliation Commission on Indian Residential Schools.* Toronto: University of Toronto Press.

Noe, Alva. 2004. *Action in Perception.* Cambridge, MA: MIT Press.

Nolan, Christopher. 2001. *Memento.* Alliance Films.

O'Keefe, John and Nadel, Lynn. 1978. *The Hippocampus as a Cognitive Map.* Oxford: Oxford University Press.

Pickles, John. 2004. *A History of Spaces: Cartographic Reason, Mapping, and the Geocoded World.* London and New York: Routledge.

Redish, A.D. and Touretzky, David. 1997. Cognitive Maps Beyond the Hippocampus. *Hippocampus.* 7(1). 15–35.

Rowlands, Mark. 2006. *Body Language: Representation in Action*. Cambridge, MA: MIT Press.

Rowlands, Mark. 2010. *The New Science of Mind: From Extended Mind to Embodied Phenomenology*. Cambridge, MA: MIT Press.

Scott, Colin. 1983. Semiotics of Material Life Among Wemindji Cree Hunters. Unpublished PhD thesis, Department of Anthropology, McGill University.

Scott, Colin (ed.). 2001. *Aboriginal Autonomy and Development in Northern Quebec*. Vancouver: UBC Press.

Sletto, Bjorn. 2014. Cartographies of Remembrance and Becoming in the Sierra de Perija, Venezuela. *Transactions of the Institute of British Geographers*. 39(3). 360–372.

Stairs, A. and Wenzel, George. 1992. "I am I and the Environment": Inuit Hunting, Community and Identity. *Journal of Indigenous Studies*. 3(2). 1–12.

Stallins, Anthony. 2011. Scale, Causality, and the New Organism-Environment Interaction. *Geoforum*. 43(3). 427–441.

Tallis, Raymond. 2011. *Aping Mankind: Neuromania, Darwinitis and the Misrepresentation of Humanity*. Durham: Acumen.

Thornton, Thomas F. 2008. *Being and Place Among the Tlingit*. Seattle: University of Washington Press.

Thornton, Thomas F. (ed). 2012. *Haa Leelk'w Has Aani Saax'u/Our Grandparents Names on the Land*. Seattle, London and Juneau: University of Washington Press/ Sealaska Institute.

Thrift, Nigel. 2007. *Non-Representational Theory: Space, Politics, Affect*. London and New York: Routledge.

Tuan, Yi-Fu. 1976/2001. *Space and Place, 25th Anniversary Edition*. Minneapolis: University of Minnesota Press.

Turkle, Sherry. 2007. *Evocative Objects: Things We Think With*. Cambridge, MA: MIT Press.

van Ham, Maarten, Hedman, Lina, Manley, David, Coulter, Rory and Östh, John. 2014. Intergenerational Transmission of Neighbourhood Poverty: An Analysis of Neighbourhood Histories of Individuals. *Transactions of the Institute of British Geographers*. 39(3). 402–417.

Wittgenstein, Ludwig. 2009. *Philosophical Investigations*, 4th edition. Chichester: Wiley-Blackwell.

Zeman, Adam. 2008. *A Portrait of the Brain*. New Haven: Yale University Press.

5   The political geography of names

The mapping of disease in Canada is political, especially mental illness and associated outbreaks such as those termed 'suicide clusters' referred to in the last chapter. As I have demonstrated elsewhere (Eades 2015), it is much easier to predict the absence of such outbreaks than to do so on the 'positive' outcomes (e.g. a suicide cluster). The absence of suicide in northern and indigenous communities in Canada can be mapped onto the presence of community-building institutions and emergency services. This is, in turn, largely tied to the political will of provincial and/or regional actors and to the historical relationships that exist in place (Chandler and Lalonde, 2009). In Quebec, treaties were not settled with the First Nations and Inuit populations until the James Bay and Northern Quebec Agreement (JBNQA) of 1975 (Gagnon and Rocher, 2002). As a result, indigenous and Inuit groups in northern Quebec have exercised a degree of autonomy and control over their internal affairs not seen in the rest of Canada, with the exception of British Columbia (Asch, 1997). This, in turn, has an effect upon the ability of local and indigenous groups to 'name' that which afflicts them in ways that respect local cultures, protocols, and styles of healing. Suicide is the name for a terrible affliction in many northern indigenous communities; it is the name for something deeper that afflicts many communities burdened with legacies of residential schooling, displacement, and colonization (Niezen, 2013). An understanding of why, how, and where suicide clusters occur is in many ways reminiscent of the state of physical disease mapping a century ago (though the analogy cannot be pushed very far), with aspects tied to outdated ideas about both the disease and its carriers (mostly the poor) and the physical mechanism of the disease (bacteria which could not, at the time, be seen but which can be seen as analogous to insidious and mobile suicide memes).

As we will see below, John Snow, MD, an early disease mapper who focused on cholera, was met by politically motivated resistance to his theory of waterborne disease. In both cases (suicide clusters and cholera outbreaks), the underlying mechanism for the outbreak was not well understood. In the absence of irrefutable evidence for a plausible mechanism, prejudice and prior ideas held sway, and these were often politically motivated. In the case of Snow, City of London public works officials had a vested interest in continuing to operate under the assumption that cholera was airborne. This assumption also had moral overtones: the poor were seen to be living in squalid conditions from which they were unable to raise themselves, which often mapped onto (at least in official minds) disease and death that many would have said the poor had brought upon themselves through inaction or laziness (i.e. the inability to find alternate water sources or raise themselves out of their condition). In the Canadian case, northern indigenous populations unfortunate enough to live in places with political legacies of inaction on reserve health often see themselves living in what can be called 'third world' conditions. For example, northern Ontario (where treaties were settled long ago) has communities that often lack basic sanitation facilities. There is an unfortunate overlap between poor physical health and poor mental health in communities without basic infrastructures, though clearly one cannot posit a straightforward correlation. What is clear, however, is that historical and political factors often define negative outcomes such as disease outbreak, lack of housing, and mental illness.

We leave Canada behind for the moment as we move next to explore the case of John Snow's London and his 'naming' of the phenomenon of cholera by way of the waterborne transmission thesis. This is an idea that has fascinated generations after Snow as the story of how he came to map cholera has (rightly) become famous. We go even further and posit counter-mapping as a form of political maneuvering in response to attempts to rename entire landscapes. In the case of Snow, he was renaming the landscape of cholera in London, the effects of which are felt down the generations in the work of doctors, epidemiologists, and others mapping the disease. The 'lesson' for northern Canada is that the earlier moment of achievement, that of Snow's identification of a pattern of disease movement and the medium through which it moved, but not the mechanism for cholera outbreak, has parallels for mental health. Though we can often trace personal tragedy retrospectively to aspects of an individual's life, extending that back into the historical tragedy of the treatment of indigenous peoples in Canada, it remains quintessentially tough to identify precisely what went wrong in a community that experiences a suicide cluster. All we know is that damaging ideas are 'out there' and that they cause negative identification as a badge of honour (Niezen, 2009). Beyond this, it could well be the case that with suicide in northern Canada it is simply too complex a phenomenon to be able to pin to one mechanism or 'final cause.' But a compelling case has been made that broken continuities and lack of connection to traditional (land-based) structures have the effect of breaking positive intergenerational memes for transmitting anchoring names on the land that in turn leads to sedentary and inactive lifestyles. Identifying areas where indigenous communities are losing ground and literally losing territory can spur community action to counteract, and counter-map, that loss of territory (Wood, 2010)

For Snow, to counter-map required the identification of areas of outbreak and so-called 'index cases', or those known to be the first instance prior to the outbreak.

John Snow and the name of cholera

In this section I posit John Snow as the pioneer of two proto-GIS operations for mapping disease. Snow is often credited with founding the discipline of epidemiology and, I would posit, it is primarily two tools, overlay and proximity, that allowed him to do so. Further, I will argue (as many others have), but in a way that makes specific use of names, that he could not have done so without physical, paper, maps (Koch, 2011; Hamlin, 2009; Hempel, 2006; Johnson, 2006; Vinten-Johansen et al., 2003; Snow, 1855). Snow is most famous for calling for the removal of the handle of a pump in Soho, London, an action that, according to myth, resulted in the instant and overnight abatement of the cholera outbreak in that area. The story is clearly mythical and elements of it do not line up with reality. In fact, cholera is not so place-bound as to be readily eradicated by such targeted action as deactivating a well (Hamlin, 2009, page 181). Removal of the handle was effective but it was viewed by many an expert as a coincidence or a lucky guess (Hempel, 2006; Johnson, 2006). The genius of Snow was his belief in and systematic gathering of evidence in support of the waterborne nature of the mechanism of transmission of cholera. Only later would that mechanism be fully fleshed out by Koch's discovery of *vibrio cholera* (which, science would later show, existed in many varieties) (Hamlin, 2009, page 211).

Snow's use of maps in conjunction with 'shoe-leather' medicine was part of a new regime of data visualization and medical techniques for establishing truth through evidence-based inquiry (Koch, 2011; Hamlin, 2009, page 157; Hempel 2006). As a result, the name of cholera has come to be associated with the place where Snow tested his groundbreaking theory, Soho. Snow's use of maps as a basis for fieldwork, and as a platform for both inductive and deductive reasoning, was elegant and powerful: it serves as an exemplar for geographical investigation and spatial analysis to this day. GIS can similarly be said to owe a lot to Snow for the two basic analytical functions mentioned above. The conjunction of proximity and overlay functions demonstrate how waterborne transmission of cholera is not simply about place (or proximity to a particular area) but is also fundamentally about space (with overlapping larger areas of interest or concern leading to focus on a specific area). Snow's creation and use of these tools was a kind of counter-mapping against the airborne hypothesis of cholera transmission that was so favored in his day. Though many of the officials responsible for the power of the airborne metaphor may not have used maps, Snow's action counts as counter-mapping because it spatially counteracted what was in effect a spatial argument. Airborne disease transmission was both a positivist assertion based on an unfounded induction of bad smells emanating from dung heaps and slaughterhouses; at the same time it was a powerful assertion by individuals with much to lose and great interest in the status quo (Hamlin, 2009, pages 153–155). We explore many of these aspects below under counter-mapping theory. For now, we explore the two analytical functions for which Snow was responsible.

The overlay function allowed Snow to make assertions about space based upon maps of supply areas of various water companies in the area in and around Soho (Snow, 1855, page 74). By establishing water as the medium of communication of cholera, Snow was going against the contagionist (or airborne) dogma of the day (Hamlin, 2009, page 180), in effect counter-mapping that dogma through the use of a map that disentangled a very complex picture of underground water flow north of the river Thames. In the second edition of Snow's most famous work, *On the Mode of Communication of Cholera*, an in-depth study of the number of cases of cholera was compared in the supply areas of two water companies: Lambeth, whose water supply came from upstream sources; and Southwark and Vauxhall, whose water came from the sewage-filled lower Thames (Hamlin, 2009, page 180). This part of Snow's experiment involved extensive fieldwork interviews in order to sort out individual addresses by asking those in residence to whom they paid their bills. Many streets were served by both companies, but specific areas to either side of the overlap were served by a single company. By eliminating other factors potentially contributing to cholera, such as proximity to bad smells, poverty, profession, and prior health conditions, Snow was able to single out water supply as underlying the mode of communication of cholera. But he did not rely on this single source of evidence alone.

The second part of Snow's experiment was his analysis, again through mapping, of an outbreak in the Golden Square part of Soho (Hamlin, 2009, page 181). The map of this outbreak was both visually and conceptually innovative and can be correctly said to consist of a sort of proto-GIS style of analysis, one that explicitly entered the thinking of those who would develop GIS software tools over a hundred years later (Chrisman, 2006). It is also a good example of what later came to be known as a map 'mashup' (Peterson, 2014). The map base showed proprietary boundaries (i.e. where people lived), with dividing lines between buildings and properties and the interstices between labeled as streets and byways. It was a black-and-white reproduction upon which Snow placed (or mashed up) two layers of locational data: well locations and incidents of cholera. The well locations were straightforward, and several were shown on the Soho map Snow used. Cholera incidents were trickier, and Snow's solution to the problem of how to visualize a large number of cases in one place was a stroke of genius.

Snow, in effect, created a 'stack' of black bars, with one bar indicating one case of cholera. The higher the stack of black bars the greater number of cases of cholera in that particular place. The striking thing about Snow's map is that the pattern of clustering around the famous well whose pump handle was later removed is very clear, though it relies for the most part upon human perception (i.e. brain power) for its effectiveness. Later GIS software tools would appear to make all the work happen in the computer, leaving the human operator to simply compile predigested "results," but they rely as much (or more) upon the spatial computing power of the human brain in order to make inductive leaps (Koch and Denike, 2006; Shiode, 2012; Smith, 2002;

Mcleod, 2000; Vandenbroucke et al., 1991; Cameron and Jones, 1983; Smith, 1982). For our purposes we will not cover the already well-trodden ground of whether the myth of Snow is justified, and the associated empirical question of whether or not he was right. Instead we look at Snow's achievement in terms of geographical naming.

Two aspects of naming become clear through a re-examination of Snow's work in light of our definition of names. First, is that the name for the proper noun *Cholera*, with waterborne connotations, arose through Snow's work. This is the ontological dimension of the name of cholera. The second is that maps provide "singular noun phrases that are used to refer to particular things" (Moore, 1993, page 1). Both (proper noun and singular noun phrases) exist as parts of NTNs for tracking the phenomenon of cholera in both its particularity and its multiplicity of reference (Hanna and Harrison, 2004). In particular, cholera can be identified, after Snow, as a singular waterborne thing that is communicated through both space and time. In terms of time, it is the evolution of the idea of cholera that is at stake, and this leads to the second point. Cholera is (we now know) a distinctly multiple phenomenon (or set of phenomena), and we now know this from modern science, that exists in multiple forms that are mappable as evidence of vectors of its spread come to light. New kinds of epidemiological maps are now easily produced, mapping varieties of DNA structure associated with particular kinds of cholera in specific areas of outbreak. Maps of this kind particularize the various strains by counter-mapping (in their own way) the work so meticulously produced earlier by Snow (Hamlin, 2009, page 255). Maps themselves can be posited to have become a sort of cultural DNA for mapping all kinds of phenomena through set kinds and types of maps for various purposes, including, for example and for our purposes, intergenerational geographical name transmission (Monmonier, 2015; Eades, 2015). We explore next how names become core parts of counter-maps, extending as well beyond the initial example of Snow used above.

Counter-mapping theory

Geographical names form a key drive at the core of the counter-mapping impulse. That impulse is, in turn, driven by politics, self-determination, and identity. Definitions and metaphors are both useful in delineating what counter-mapping is. Said (1993, page 51; quoted in Sparke, 2005, page 4) has used a mapping metaphor in defining post-colonial sensibilities as "contrapuntal cartographies" that indicate "a simultaneous awareness both of the metropolitan history that is narrated and of those other histories against which and together with which the dominating discourse acts."

For Snow, a Yorkshireman of humble means, working in London meant working against the dominating discourse of those in power, and those with vested interests in maintaining a status quo contagionist and miasmatic (airborne) view of cholera transmission. Snow's ability to rename the playing field through the use of maps, empirical investigation, and fieldwork thus enabled a contrapuntal cartography and counter-mapping that would take many decades to be seen as a break from the previous paradigm.

The definitional approach to counter-mapping has been best described by Peluso (1995, page 384), though it falls short of a true definition, remaining as a suggestion as to counter-mapping method: "the goal of these efforts is to appropriate the state's techniques and manner of representation to bolster the legitimacy of 'customary' claims to resources." For Snow, this meant the use of a proprietary base (property) map upon which an overlay of separate information could be used for analysis, including wells and raw counts of incidents of cholera. Though Peluso was speaking above about indigenous counter-mapping efforts in a modern world of multi-national forestry, the spirit of the definition does not elude what Snow was doing in mid-1800s London. For it was, ultimately, resources that were at stake, and those resources were primarily in the hands of those in thrall to the state—state-like, or state-influenced, actors and interests. Snow, for his part, can be construed as a pioneer, a non-conformist, or a rebel, depending on the narrative demands of the story being told. For our purposes we wish to posit Snow as not only a proto-GIS thinker but also as an early counter-mapping anti-hegemonist (or anti-positivist) (Hamlin, 2009).

How do names fit into all of this? Moving away from Snow, it is very clear that toponyms, or place names, form a key component of counter-mapping, as the struggle for land rights in the face of state or multi-national incursion usually revolves around demonstrating intergenerational presence and resource use on the lands in question prior to colonial incursion or interest from outside. *Location, scale, density,* spatial and temporal *extent, language* type (and the related issue of *meaning), continuity,* and *practice* (both as engaged with and as used) are all at play, especially when courts become involved in arbitrating indigenous interests against hegemonic ones, through the use of maps. These attributes of counter-maps are discussed below.

In a discussion focused mostly on England, Hanna and Harrison (2004, page 104) shed some light on what is at stake in 'indigenous' (to England) naming (and by extension counter-mapping) practices, and the establishment of the continuity or validity of reference through time, referring themselves to the sketchy place name 'Easthampton,' which today cannot be found with any precision on any authoritative (UK) map:

> What is it that in practice permits the members of the historical community to credit themselves with the ability to refer to Easthampton, and to attach assertoric content to sentences in which the name occurs, even though no member of that community has the faintest idea which, of a number of areas of early mediaeval architectural vestiges, occupying widely separated sites, are the remains of Easthampton? Given the textual character of historical data, those abilities can only repose on the possession of what we have been calling a nominal description of Easthampton, or rather a

series of such descriptions. Easthampton is known to historians, let us sup-pose, as the village of that name referred to successively in the Domesday survey, a sequence of characters, and a range of surviving letters and legal documents relating to disputes of the thirteenth and fourteenth centuries over feudal rights relating to the village.

(italics in original)

The authors (Hanna and Harrison, 2004, page 105) go on to describe why precision in language is so important for the study of place-names as linguistic artefacts and as parts of place-naming practices:

> Although our use of place-names is sometimes quite loose (where, for instance, do 'Middle America' or 'The Levant' begin and end?), we go to a great deal of trouble to establish and operate far stricter criteria of identity for named 'places' falling into a range that includes cities, towns, villages, districts, fields, commons, gardens ('messuages'), estates, and others. We establish the boundaries of such entities with care, marking them with boundary-stones, surveying them in order that they can be accurately delineated on cadastral maps, town plans, estate maps, and so on. We do these things because they serve, are indeed essential to a wide variety of purposes arising in the context of a wide variety of legal, commercial and governmental transactions; transactions involving inheritance, conveyan-cing, the areas of responsibility of adjacent local authorities or police forces, feudal rights and duties, planning regulations, and so on. A place-name of this type, a name such as 'Easthampton,' or 'Baxter's Piece,' or 'Middle Farm,' once it gains currency, rapidly begins to acquire … a use [in practice], or rather a whole fan of related uses [in practice].

The 'scare quotes' around the word 'places' are telling. Places are too easily associated with names in their proper sense. We tend to forget that unnamed places exist and that still (paradoxically) consist of names in the sense of noun phrases in language that refers to particular things in the world: walls, road intersections, clusters of buildings, fields, and many others. Parts of boundaries can be seen as unnamed places in the proper sense, but they are very much named (geographically) both in terms of language for referring to objects in space and relationally in terms of spatial relationships *between* the objects. Medieval and modern day rogation practices make this abundantly clear, as seen in chapter 3. The idea of attaching "assertoric content to sentences in which the name occurs" (Hanna and Harrison, 2004, page 105) is relevant here for elucidating one of the present work's most forceful points, that geo-graphical names, place names, and toponyms do not come close to exhausting meaningful senses of place, belonging, identity, and power that are important parts of culture and its evolution. Yet both the geographical literature and popular notions of geography often assume a language of names whose dis-course is made to appear neutral. The result is often erasure of naming practices that lie beyond the ken of ordinary naming or use of language. We seek here to map anew extraordinary, relational, unconscious, and power-laden aspects of language, naming, and space.

In this sense the present work is also a counter-mapping against both dominant discourses in geography that assume names in facile or unproblematized ways and popular views of geography as rote memorization of place names, geographical facts, and GIS software routines. Mapping, in this book, is more metaphorical than literal since we use primarily language and new ways of conceptualizing geographical space to elucidate a new space for geographical names, naming systems, and practices. At the same time, this approach does not diminish how maps are used as powerful tools for both visualizing and representing geographical names in space. The top-down view that maps allow is as productive of accumulated meaning (over time) as the grounded, phenomenological view. Indigenous groups in northern Canada epitomize a dichotomy between view-from-above and grounded views with their sophisticated use of both on a daily basis, especially in active hunting-gathering-fishing communities. As this author has often noted, the best way to start a discussion in a Cree or Inuit community is to pull out a topographic map.

The magic of maps

Maps are essential tools in NTNs for tracking not only names over time but also relationships between geographical objects, political entities, areas, water-bodies, properties, identities, and many other natural and human phenomena, features, and events. The salient point here is the map's status as a tool, but we wish to point out that maps are part of a necessary apparatus for representing space objectively, from above, and in as neutral a manner as possible. Many authors have pointed out that maps are socially constructed objects of power and persuasion, and this assessment is surely correct (Dunlop, 2015; Husain, 2014; Padron, 2004; Edney, 1997). But as Nagel (1986) has pointed out, maps (or, more properly, map-like ways of seeing the world schematized, 'from above' and, metaphorically, as 'from nowhere') are also tools for making sense of objective–subjective dichotomies that complicate the human condition and that, Nagel argues, have driven one of the most interesting problems in philosophy. Therefore, in this book, while I accept social constructivist theories as fact, I also accept and problematize representationalism as essential to any theory of geographical names and maps-as-tools for tracking them through time.

We argue therefore, that maps are parts of more-than-representational networks that in part track names but also do work in the world. In this sense, as noted above, this book aligns quite nicely with ANT (Latour, 2005). The visual nature of maps means that it is easy to accept them uncritically as true whether or not what they depict is true in an objective and/or subjective sense. This is equally true for the printed word, perhaps more so, but maps are always either partially or fully realized as texts, consisting of words and images, with the drawing or 'data' forming a bridge between the two (God-lewska, 1995). I am focused here more on the language side of the mapping 'bridge', but this should not diminish the power of the 'other' side of maps, the drawing and imagistic power that makes them so useful and persuasive as tools for producing meaning and, objectively, for tracking named places through time. The image can be seen as a supplement to the word (the name) for identifying places that may not always be named properly or that form parts of boundaries, boundary objects, or relationalities inherent to particular places.

Edney (1997, page 40) has noted that geographers in particular have often

> represented the structured assemblage of geographic knowledge as a concrete structure in and of itself. This was not an uncommon strategy. The built environment of the archive and the museum has long served as a fundamental metaphor for modern European conceptions of knowledge creation. Data and artifacts can be collected within sturdy walls and there reassembled into meaningful arrangements. Indeed, the walls are overly protective. They physically divorce the collected data and artifacts from the actual contexts of their occurrence and existence.

In terms of tracking names, in large part through the use of maps, I argue here that this metaphor is not misplaced. I explore several issues related to counter-mapping below in light of what can be adopted and co-opted from the European conceptions of mapmaking to support indigenous and local naming systems and the preservation of their continuity through time. This activist sensibility is justified by the globalizing pressure of new mapping technologies, neo-colonial mentalities (Gregory, 2004), and renewed pressures for local groups to both conform to and adopt dominant geographical naming strategies.

Attributes of counter-maps

Precise *location* of geographical names is of great importance in counter-mapping efforts—for example, in the 'contest' between the Gitksan and the Wet'suwet'en against the Nisga'a, whose competing claim necessitated the court proceedings alluded to in Sparke's paper, "A Map That Roared" (Sparke, 2005). But in this challenge (and others), precision of location is not the be-all and end-all of the case (i.e. it does not make or break it). Named geographical entities show considerable variation depending on the nature of the feature and whether it generally conforms to a point, line, area (i.e. vector), or even field structure (raster). In this case, the evidence was both oral and textual—recounted from memory by elders in Gitksan and Wet'suwet'en communities, with subsequent inscription onto western-style cartographic products.

Scale is very important for all of the following reasons. First, it affects the number of locations that can be shown on a map. For this reason, for example, the Inuit Land Use and Occupancy project resulted in an entire atlas of counter-maps, generated through interviews with elders knowledgeable about the northern terrain through hunting activity (Bryan and Wood, 2015). One map would not have been sufficient for the purpose of the Inuit living in what is now known as Nunavik. The latter is a vast portion of northern Canada, one of its largest territories, and to represent the entire area on one map would have precluded a density of names required to assert the validity of the claim (Brody, 1975).

In terms of *density*, the fact of it is alone a direct refutation of colonial-era and imperial maps that depict space in the new world as essentially empty (Akerman, 2008). Harley's (1989) seminal paper "Deconstructing the Map" was in large part a debunking of the idea of empty space as anything other than a social construction (see also Brealey, 1995; Harris, 2002). Space is never just an empty neutral or static backdrop for phenomena in the world. Instead, it is always already made by something or someone (Lefebvre, 1991). Indigenous spaces and places were being made for decades, centuries, and even millennia before the arrival of European explorers, traders, and surveyors (Lewis, 1998). Through the production and dissemination of counter-maps, created in the lands and resources offices of indigenous peoples themselves, cartographic errors (often produced by the blinders of colonial or outsider cartographic power) could be corrected and displayed for consumption and used as evidence in courtrooms (Eades, 2015; Ray, 2012).

At the same time as density is defined through extensive interviewing, field survey, and other empirical verification techniques, *extent* of territorial boundaries will gradually become clear. Extent also works in conjunction with language types, described below, and elucidated so well in the Alsace-Lorraine dispute described by Dunlop (2015), in which amateur cartographers took it upon themselves to map local areas in German or French, according to the interests and alliances of the individual cartographer. Brody's (1981) classic text *Maps and Dreams* presented a method of defining territorial extents for use on counter-maps intended for governmental surveys and archives. Individual hunter territories were defined through sketch maps, which were essentially 'loops' or 'circles' defining how far each hunter would go in search of game. The individual maps were overlain in stacks atop each other, producing a chaotic looking 'scribble' map, the furthest extent of which defined the collective edges of the territories (in the case of Brody, for the Dene people of northeastern British Columbia, Canada).

Language type and the related attribute of *meaning* are both important in establishing boundaries. As a bare minimum, the toponymist must record the location of a name with x and y (e.g. latitude and longitude) coordinates (an absolute method for locating) or through some relative means such as proximity to other features, or length along a traveled path from some other known location. The toponymist must also record the name of the place in whatever language it is currently named and, along with this information, the meaning of that name must also be recorded, as well as the name of the individual (e.g. elder) from whom the information was ultimately derived (i.e. the 'informant') (Kadmon, 2000). Without meaning (which in turn requires translation) the name remains a mere tag or label, without the possibility of ascertaining the encyclopedic knowledge (i.e. stories, facts, or resources) associated with it. Language type is also important for the ascription (and inscription) of boundaries between different spoken languages that can, in turn, define the extent of the area of territory claimed by the counter-map to fall within the remit of the mapping agency or individual. The Alsace-Lorraine region between France and Germany is a case in point. This region is defined not only by physical features, with the Vosges mountains on the French side and the Rhine River on the other, but running down the middle is a so-called language line, on either side of which (in theory) different languages are spoken. In practice, on the ground, language is a gradation or a continuity between French and German, with subtleties in the spoken word changing as one moves from west to east—for example, from France into Germany (Dunlop, 2015).

Continuity is difficult, if not impossible, to positively assert, but it is implicit in many land claims discussions, especially in Canada. If a current place name represents a break from the past, then continuity must exist at some other level (i.e. in practice). Margaret Gelling's life work was devoted to tracing linguistic threads through time by tracking word particles (parts of words) in order to locate and elucidate meanings of names across space, and through time, in England and Wales (Gelling, 1978). Gelling points out that place names in the UK can be derived through iterations and translations of Celtic, Roman, British, Norman, Scandinavian, Anglo-Saxon, Old, Middle, and Modern English. The possibility for mistranslation or misappropriation of place names is great and requires rigorous scholarship and subtle understanding of language, geography, and history (Padel and Parsons, 2008). Geographical names can, however, be translated or tracked through their situation in land-scape. The contextualization of toponymy in this way, by associating meaning with geographical features, is of great importance in establishing continuity of naming practices and also human and physical worlds in culture (Gelling and Cole, 2000).

Last, we discuss *practice* as both engaged with and used. Naming practices form bridges as language (the word or the name) links into practice and practice in turn is used in the world. Mapping is a practice that forms scaffolding for supporting an overlapping concern with naming, as well as other forms of representation (including visualization) with which the map has an associated or demonstrated strength. The idea of practice forms a core argument of the present work. Counter-mapping is both caught up in the logic of geographical naming practice as political activity and a necessary counterpart to struggles to assert a multitude of competing identities in an ever-changing and evolving world. The risk in such a world is that long-inhabited and, in some cases, fought-for territories with which those identities and struggles are associated will somehow become subject to erasure or catastrophic change such that both the material (base) and super-structural elements of culture suffer or disappear altogether. Practices for naming and mapping the land, on the other hand, seek to (re)inscribe territorial difference for long-term storage in institutional and individual memories.

A map can exhibit all of the above properties of having clearly defined scale, extent, density, locations, and meanings mapped without being a counter-map. In fact, one could argue that any good map would have such attributes. A disaster map (Tomaszewski, 2015) must clearly have a defined extent of the disaster, such as an earthquake, and show the density of mapped locations with interpretation of their significance (low, medium, or high levels of risk) on the ground, and so on, and the database behind the map must be linked with names and associated attributes for the map or GIS to be useful. But such a map would not be inherently political in the same way (i.e. not as straightforwardly) as a counter-map (Wood, 2010). Counter-maps wear their purpose on their sleeve, so to speak, by including national logos, or by weaving flagged colors into the design, or simply by including images of local resources and/or people. Not all counter-maps will exhibit all of these attributes, but most will likely include some. New kinds of artistic assemblages and psycho-geographical mappings are coming to be collectively 'filed' under the auspices of counter-mapping, but many of these would eschew notions of continuity, instead offering a vision that breaks, often quite radically, from the past (Self and Steadman, 2007). For our purposes, however, we focus here on counter-mappings that assemble elements in new ways for reconfiguring indigeneity against dominant (often state-based) power structures, naming systems, and practices (Eades and Zheng, 2014). There is a long history of this kind of counter-mapping in Canada, in both long- and short-term time scales, of indigenous peoples' defining struggles for self-determination, identity, and continuity of culture against colonizing forces that seek instead erasure, broken heritage, and assimilation.

The point of counter-mapping is that maps made by outsiders often misrepresent local interests, either intentionally or through benign neglect and/or error. Early mapping of Wales often took place in the Netherlands and, a direct result of this kind of 'outsider' mapping of the local, place names were misspelled, distorted, or dropped altogether (North, 1935). In the case of Canada, it was more often outright erasure of indigenous geographical and personal names that was part and parcel of a kind of blank-slate mentality that drove the colonial impulse (Agamben, 1998; Eades, 2015).

Counter-mapping Canada's north

Canadian indigenous peoples, when they are noticed at all, are often best known among mainstream media as forming protest movements (Belanger and Lackenbauer, 2014). Canada's indigenous inhabitants are generally seen as radically 'other', inhabitants of waste spaces, interstices, and undesirable places. Conflict arises when valuable territory is perceived as occupied by one side or the other (i.e. by a First Nations group, on the one hand, and an interest of the Canadian state, industry, or dominant society group on the other). Simple recognition of indigenous culture as viable (in the sense of alive and continuing to evolve) and valid (in the sense of representing sophistication contemporaneous alongside mainstream Canadian culture) is a massive challenge. But this step, of establishing a core of legitimacy in the eyes of court and media actors (not to mention the general public), is a very large one, and necessary for indigenous struggles whose goal is self-determination. Counter-mapping, as part of this legitimating move, often uses culture in the form of art, icons, and logos in order to draw the reader of the map in with something easier to identify than, for example, a transliterated Cree place name. While the latter (geographical names) represent a core component of the counter-map in all its density and clarity of extent, it is the former, the art or the logo, that might draw the reader in initially and result in identification with what the map depicts.

Other counter-mapping moves used by First Nations groups in Canada include modifying traditional cartographic elements in subtle ways for promoting a message of territorial integrity. A north arrow or scale bar can be embellished with the forms or colors of, for example, northwest coast art or the flag for a particular region. Naming, itself, is, however, the fundamental move in First Nations and Inuit mapping of Canada's north, and here we must back up a bit and explore the foundational place of the proper names of individuals. Unlike in non-indigenous cultures, indigenous groups do not have a tradition of naming places after people (Monmonier, 2006). Several authors have noted the importance of names to the Inuit of northern Canada, collectively exploring how continuity over time is maintained in Inuit naming practices (Wright, 2014; Alia, 2007; Bennett and Rowley, 2004).

Continuity in the Inuit NTN, for example, is maintained by way of what is called (in Inuit) a 'bone' or *sauniq*. Wright (2014, page 39) notes that, "[p]eople who share the same atiq (name) are of the same 'bone,' the same spirit, and have a special relationship." She goes on to describe how continuity of names work in greater detail:

> Usually a name is passed from someone who is already deceased or close to death. A name may be shared among many people, all of whom also share a relationship with each other. When Inuit are close to death, they often ask that their name be passed to a newborn baby or to a child about to be born. Names need not be of relations but may also come from friends. The passing on of names has occurred over many generations for hundreds, perhaps thousands, of years. Each time a name is passed from one person to the next, the spirit belonging to the name passes with it.
>
> (Wright, 2014, page 39)

It is all the more tragic, therefore, that the Canadian government chose to treat personal Inuit names in much the same way as they treated geographical names on maps. This was achieved by three means: *replacement, erasure*, and *displacement*.

In the mid-1900s the Canadian government *replaced* Inuit names with numbers, going as far as issuing dog tags to make the numbers 'stick' better to their intended 'referents' (Inuit individuals). Many Inuit elders still remember their numbers and have retained the metal tags or handed them on to their descendants. The practice of *erasing* identity by replacing names with numbers has been well documented among Jewish survivors of concentration camps. The replacement of a personal, proper, name with a number is an act of obliteration. Hanna and Harrison (2004, page 153) make a very salient point through a reading of Primo Levi's written experiences of Auschwitz in his book *If This is a Man*:

> The difficulty for those who see no difference between baptism and labeling, is to explain why the point is so insisted on: why the mere lapsing from use of a name should be felt as entailing a loss of identity. After all, when he felt these things Levi was still alive. Here he is in the camp: Levi the man. He still has his identity because he *is* his identity. He cannot lose that unless and until he is killed. How, then, can it make any difference to his identity as an individual human being whether that identity is labeled 'Primo Levi' or '174517'?

Indeed, the Inuit of Canada have faced similar existential threats and questions over the decades of the ascendancy of the Canadian state. In order to assert territorial sovereignty, many Inuit were *displaced* to northern locations to 'beef up' the numbers across the north in an attempt to portray even inhabitation to potential competitors (e.g. Russia or the US) for territorial and geopolitical space. The assumption here was that Inuit are Inuit no matter what the location, that hunting, fishing, and other activities can be transplanted to any Arctic locale regardless of reduced day length or harsher environmental conditions in those new locations compared to the individual Inuit's true home, where they and their ancestors were raised and learnt to interact with the specific landscapes of that home. For example, some Inuit from Ivujivik were sent to northern Ellesmere Island, thousands of kilometres away and an alien and much harsher and more inhospitable environment.

Inuit were also given European-style surnames in an attempt to assimilate them to mainstream culture (Alia, 2007). This goes against Inuit practice, where the

> tradition of passing down names extends to the nature of the relationship between the deceased and others in the family or community. If a child is named after his grandmother he becomes in a way his own grandmother. Other Inuit will refer to him according to their own relationship with the deceased grandmother. Since Inuit often refer to each other by the relationship they have with each other (as calling someone by his or her given

name can seem disrespectful or may even be forbidden), a baby boy may find himself being called *anaana* (mother) by his own parent, aunts, and uncles—everyone who was a child of his grandmother!

(Wright, 2014, page 40)

We argue here that the same operations of replacement, erasure, and displacement that were performed upon Inuit personal names were also applied to their geographical names. Indeed, personal and geographical names can be posited as belonging to a single overarching, NTN-related continuity of Inuit culture in its entirety. This NTN is made up of much documentation and inscribing devices such as dog tags and maps for performing the operations and using the colonizing tools for the displacement of Inuit culture. Furthermore, the strategy applies to First Nations and other indigenous cultures across Canada. Strategies for resisting and overcoming these damaging legacies can take similar forms across time and space. They include reclaiming, through re-inscription and performance (bodily or otherwise, such as storytelling and journeying), traditional names as part of rebuilding and asserting the continuity of the indigenous NTN. This often takes the form of counter-mapping, as described above and elsewhere (Eades and Zheng, 2014).

Eponymy

Inuit naming systems make use of eponymy to ensure continuity in their traditional NTNs that have existed since time immemorial. We might know eponymy best from Eurocentric naming systems that preserve the last, not given, name of the father. We might also know about eponymy, which is after all a fairly unfamiliar word even in place-name studies, from the practice of assigning names from the 'old world' in the 'new world.' Examples of this kind of eponymous naming abound in North America: for example, in Ontario we have a city named London and a river named the Thames within a couple of hours' driving distance from Lake Ontario and Detroit. Eponymy serves to commemorate the old place in the new, but it is not necessarily based upon resemblance, except perhaps subjectively, in the nostalgic mind of the beholder/colonizer. North American indigenous geographical names do not rely upon eponymy, though we cannot know this with 100 percent certainty. After all, the Inuit of northern Canada do not consider themselves indigenous to the Canadian Arctic, having displaced and interbred with the earlier Dorset (palaeo-Eskimo) people in what is known as the Thule migration from places further west (Dorais, 2010 and 1997; Laugrand and Oosten, 2010). It is possible that eponymous geographical naming has occurred in the north, though due to the descriptive nature of indigenous geographical names it is somewhat unlikely. More likely is that names follow different categories of use value, such as those for transportation (along rivers, lakes, or ocean inlets and passageways), for resource harvesting, mythical and real events, stories, and dwelling places (Eades, 2012; Afable and Beeler, 1997).

Unlike in Euro-American systems, eponymy is used for the given names of Inuit individuals, and this eponymy often has a spatial component linked to the land- and sea-focused life traditionally followed by the Inuit. The *uqalur-ait* directional snow dunes offer a metaphor for making sense of names (geographical and personal) in Inuit worlds. According to an informant of Bennett and Rowley (2004, pages 115–116) named Abraham Ulayuruluk,

> [d]uring a blizzard the snowfall is usually soft. A type of snow mound, *uluangnaq*, is formed. The [prevailing wind] then erodes this mound thereby forming an *uqaluraq*—a drift with a tip that resembles a tongue (uqaq)—which is pointed and elevated from the ground. From this formation another kind of drift we call *qimugjuk* is going to build up. *Uqalurait* are formed by the *uangnaq* (west-northwest wind). There is very minimal formation of *uqalurait* made by *nigiq* (east-southeast wind) and these are usually very small ... . I believe the *uangnaq* is the strongest of all the winds that cause *uqalurait* except for *nigiq* which can also cause *qimugjuit* on the lee side of rocks. On a smooth surface, without rocks or pressure ridges, *qimugjuit* can only be formed by the west-northwest wind. Some of these drifts remain *qimugjuit* while others will be transformed into *uqalurait*.

The personal names of Inuit are, metaphorically, directional indicators for cultural continuity and the integrity of their NTNs. They are, in other words, tools or memes for making sense of both time and space and for structuring both in the belief systems of Inuit life-worlds. It was noted by Mark Ijjangiak, another informant of Bennett and Rowley (2004, page 6), that his namesake (the ancestor from whose name his own given name was derived) granted him special powers when traveling on the land:

> My father grew up with ... Pauktuut, as he was adopted by her. One spring, when he was a boy, they were on their way back to Tununirusiq from Tununiq, following the coast. Travel was difficult because of the thaw. Soon they came upon open water at a river, and as they could no longer continue, they just stayed on the sled. Suddenly they heard someone singing from the direction of the land ... . My father's adoptive mother knew it as the place where my father's namesake, her grandmother, had died and was buried ... . The adoptive mother recognized the voice—it was my father's namesake. At once she said to my father: "*Atiruluit inna tikittatukalaurlavung. Atiin Inna tusaqsauvug* [Let us go and drop in to see your namesake. It is your namesake who is singing]."
>
> They went up to the land and came upon the grave, which was made with stones. The two went around the grave clockwise, then stopped where they had started, and then made another two rounds and stood for a while. There was of course no evidence of anyone capable of singing in this untouched grave. After a while they returned to their sled just to discover that they now had a route to take to continue their journey.

    This particular incident was interpreted as the namesake wanting to meet the name bearer. That is what I have heard anyway. When my father reached adulthood, though he did not get to become a shaman, he nonetheless had the power to pinpoint the location of missing people. When someone was lost, he would be consulted, and would say where he felt the searchers should look. He was usually right. It is said he was able to do this with the help of his namesake

Just as Christianity is coming to overlay and create hybrid religious structures with the older shamanistic religion of the Inuit, naming systems overlay each other. The informants above have obviously taken European-style (first) names, which has become a new tradition in the Arctic. But the old namesakes and linking names are still used amongst themselves, preserving the old ways of kinship, community, and land-based dwelling alongside the newer ways of naming, being, and dwelling on the land. Retaining the old ways can be seen as a kind of counter-mapping-as-naming assemblage for the creation of newer, more complex Inuit and indigenous identities in the Canadian north.

    Many other aspects of Canada's geographical (as opposed to metaphorical) north are becoming reconfigured alongside indigenous and Inuit cultures. The North Pole and the Northwest Passage are just two examples, and these are briefly explored below.

Deconstructing the North (Pole)

What nationality is Santa? This question is more serious than it sounds, cutting towards geopolitical tensions created by the idea of nations 'owning' or 'claiming' the North Pole as their own. The North Pole is a very complex place, despite what it might look like 'on the ground.' In fact, even that is a misrepresentation, as the pole's manifestation at the earth's surface is on a patch of the frozen Arctic, a seeming wasteland of white. Beneath that picture of a white, featureless plain, the image seethes with contestation and deep contradiction. There are at least five different ways of seeing the North Pole: as *magnetic* (Merrill, 2010), as *geographical*, as *metaphor* (Donaldson, 2014), as *cultural construct* (Baldwin et al., 2011; Bennett and Rowley, 2004), and as *political entity* (Smith, 2014). As we explore each of these in turn it will become necessary to question and critique the very idea of 'north' (Gould, 1967) that lies at the core of the North Pole construct. The latter (the pole) is only part of what it means to be in the north (i.e. closer to the pole), and is not its essence.

    Lackenbauer (2013, page 1), writing about the Canadian paramilitary force known as the Canadian Rangers, predominantly Inuit and indigenous citizen-soldiers who patrol the Canadian north, describes the following scene:

    Cape Isachsen, Ellef Ringes Island, Nunavut, longitude 78°8' N, latitude 103°6' W, 18 April 2002. It was a biting minus thirty-six degrees Celsius with wind chill ... . Operation Kigiliqaqvik Ranger, named after the Inuktitut word for 'the place at the edge of known land,' covered more than 1,600 kilometres of rough sea ice, pressure ridges, rocky river valleys, and breathtaking expanses of tundra ... the patrol had travelled more than 800 kilometres when it was forced to stop on the sea ice north of Ellef Ringes Island. Two kilometres ahead lay a huge, impassable lead—a crack in the sea ice over 400 kilometres long and 5 kilometres wide. By attaining 79°N latitude, the expedition technically could claim that it had reached the magnetic pole—the point where the Earth's magnetic field points vertically downwards, 'wobbling' in an oval up to 200 kilometres in a single day.

It is clear from the above passage that magnetic north is neither coincident with geographical north nor is it a static feature, ranging around a wide area within a short space of time. The Inuit word for the area, *kigiliqaqvik*, is also only a very approximate name for the place at which the patrol arrived on April 18, 2002. Merrill (2010) describes how magnetic north has actually reversed several times over a vast time scale. The use of magnetic north for orienteering is a well-known but more recent development for the Inuit, who, as noted earlier in this chapter, use the more reliable *uqalurait* (directional snow drifts) for ascertaining direction.

The North Pole was not always thought of as an especially important place, lying as it did outside the *oikumene* (inhabited area), beyond the area known as Thule in Ptolemy's day (Berggren and Jones, 2000). The idea of Thule as the most northerly inhabited place, and an island, was put forward by Pytheas but refuted by Strabo (Strabo, 1917, page 441). A set of four polar islands divided by rivers were posited as real by Mercator (who was possibly influenced by the early Greek ideas of the northernmost islands) through his inclusion of the (mythical) islands in his polar maps (Crane, 2002). It is Mercator, however, who is most responsible for the idea of the North Pole as an empirical (numerical) fact, due to the ubiquity of his projection in early atlases and on maps that continue to grace our walls and books today. Mercator's eponymous projection distorts the true extent of the north by stretching the polar regions beyond their true areal proportion, greatly magnifying (and misrepresenting) the sheer extent of northern areas in relation to the rest of the globe. Thus, Greenland will be represented on a map produced using the Mercator projection as being larger than Africa despite the fact that in reality it would fit into Africa ten times over.

But clearly the North Pole itself is merely that, a pole, which is in effect a point when looked at 'top down' on a map. However, the mathematics of the Mercator projection dictates that that very same point be stretched across the top of any map made using this projection, making it equal in extent to the Equator stretching across its center. Clearly such a representation of the pole is absurd, but in the absence of more sophisticated projections it has persisted for many decades and, indeed, centuries (Arlinghaus and Kerski,

2014). Easy as it would seem to correct the situation by asserting (either by stating or mapping) that the North Pole is simply a point, cultural and geo-political realities dictate quite clearly that it is not (only) a point but can be conceived and displayed using a variety of geographic primitives known in GIS as points, lines, areas (polygons), and volumes (Schuurman, 2004). In everyday language, we can speak of the dimensions of the North Pole as being 0- (point), 1- (line), 2- (area), and 3- (volume) dimensional, depending on how one chooses to represent it. The North Pole can be conceptualized in all these ways and this perhaps helps suggest best of all what a non-place the North Pole really is. Like an airport waiting area, one does not stay any longer than one needs to (Auge, 2009). The place, according to this construct, has no meaning beyond being between two other places, for example, Russia and Canada. The myth, of course, for children, is that Santa Klaus lives at the North Pole and that his reindeer have magical powers, and this is certainly part of the meaning of the North Pole, but more in the realm of fiction (see chapter 7) and belief (the subject of chapter 3). The geographical name, North Pole, thus refers twice: once to the 'real' North Pole that we can see and touch and travel to; second, to the fictional place of stories and Christmas (on the problem of so-called 'empty' reference, existence, and meaning, see Kripke, 2011 and 2013). Atlantis as a fictional and mythical place can be treated similarly (Plato, 2008), though its existence is clearly more questionable.

Taking the North Pole as a set of dimensions, we now cover each in turn. The 0-dimensional view of the North Pole, as noted above, is the map view, and it is the most common. A pole-centric projection, one that centers the map on the North Pole, will represent it as a point. Clearly, it is a very special kind of point—the geographical extreme where latitude is 90N and longitude doesn't apply (due to the convergence of these lines at the pole). But the dominant, even hegemonic, view of the pole (i.e. the 0-dimensional) does it great disservice and offers the most to overcome in terms of thinking about polar representation and politics. On the other hand, this dominant view does the greatest service to a country like Canada that subscribes to the so-called sector territorial principal enabling it to claim a wedge of land between 60 and 141W longitude, with the former meridian defaulting to the Davis Strait after passing between Greenland and Ellesmere Island (Smith, 2014, page 192). The sector principle works in Canada's favor as it extends to the North Pole and includes the entirety of the Northwest Passage. It is disputed by the United States, which maintains that the Northwest Passage is international waters, and the US claim is ever more urgent with the opening of those waters due to climate change and Arctic warming. It is Russia, however, that most disputes the sector principle's claim to the North Pole.

The Independent newspaper ran a story on August 5, 2015 entitled "Russia submits new claim to swathe of Arctic rich in oil and gas." It is here that we begin to move beyond the 0-dimensional (point-based) representation. A 1-dimensional view of the North Pole would see its vertical extension both above and below the surface of the water upon which it lies in the map (0-D) view. Below lies the Lomonosov Ridge claimed by Russia. As noted by *The Independent*'s Isachenkov (2015, page 24),

> [u]nder international law, a country can claim exclusive economic rights over the continental shelf up to a 200-nautical mile limit off its coast. Russia says an underwater mountain range known as the Lomonosov Ridge, which stretches across the Arctic Sea, is part of its own Eurasian landmass, pushing its claim past the 200-nautical mile limit … . In 2007, Moscow staked a symbolic claim to the Arctic seabed by dropping a canister containing the Russian flag on the ocean floor from a submarine at the North Pole.

This idea of 'marking' the pole, or indeed any other 'discovered' place or piece of territory, in the name of a country has a long history, especially in the Arctic, where, as seen above, such styles of claiming are alive and well. In past times, the name of the place so claimed was inextricably linked to the name of the leader under whose auspices a country explored, colonized, and conquered new spaces well beyond existing national territorial limits. For example, in the name of the Dominion of Canada, King Edward the VII took possession of Ellesmere Island in 1904 through proclamation and the planting of a flag at Cape Sabine. The name of this royal figure, the king, claiming territory for Canada, the former British colony, was inscribed into the proclamation and linked territorially to Ellesmere Island by way of various representations, including the proclamation itself, the physical location and very existence of the flag, and photos taken of the team of explorers responsible for the 'discovery'. As noted by Smith (2014, page 148), a copy of the proclamation was sealed and deposited in a metal box inside a cairn at the location.

The exercise of power at a distance from an imperial center is a key feature of geopolitical activity of nation states (Dodds, 2014; Agnew, 2003; O'Tuathail and Dalby, 1998). The historic claims made in the name of King Edward VII at Ellesmere Island and that of Vladimir Putin on the seabed at the Lomonosov Ridge are no different in this regard. But there are distinct two- and three-dimensional aspects to the case of the current claim to the North Pole that is different from the claim for Ellesmere Island in 1904. The areal (2-D) view would represent the North Pole as a distinct region of the Arctic that exists between, and in tension with, several nation states, including not only the US, Canada, and Russia but also (at least) Norway and Denmark. There are clearly actions at a distance to the 'main' territorial boundaries of each of these nations, and as such, the attempt to exert power over those distant spaces represents a classic domain of geopolitics. The fact that huge reserves of oil and gas are at stake only makes it more so.

The volumetric (3-D) view of the North Pole is the most complex but also the most realistic in some ways. In fact, the 3-D view is alluded to in the 1-D insofar as the point-based pole representation is projected downward into the continental shelf claimed by Russia. This kind of projection, imagined perhaps as a 3D animation or 'tiltable' globe with translucent layers, presupposes the vertical nature of a territoriality dimension to human geographical inquiry that is gaining hold in the discipline of geography and reorienting it in some ways (Adey, 2010; Braun, 2000). A similar disciplinary reorientation has occurred around conceptualizing oceans as extraterritorial entities claimed in various complex ways by nation states seeking to exert power over space in competition with each other (Steinberg, 2001). The naming of aerial, marine, and other extraterritorial spaces is correspondingly complex, reflecting what geographical naming would otherwise seem to simplify. Geographical objects and ideas require names commensurate with their complexity, but they also require indexical referencing systems that inherently simplify. Thus we have boundary-making practices that invoke noun phrases as tools for dealing with multi-dimensional complexities of property, political contiguity, and inclusions of various sorts. Sovereignty over space is neither simple nor straightforward. At the end of the day, however, proper names often win out (i.e. over noun phrases) as the final 'product' of various language games played out over boundaries and exclusions that come to define nation states. With that said, the proper name (e.g. of Canada) is usually an extreme condensation of struggle, inscription, boundary description, ideology, colonization, and many other aspects (Smith, 2014).

The Northwest Passage is an evolving entity with a complex history, as entities referred to variously as Canada, the Inuit, proto-Inuit cultures, and states compete for the use of the space for economic benefit. The United States has a huge interest in revising the status quo Canadian view (the sector principle) towards the international waters vision, in which passage across northern Canada would be much the same as crossing the Atlantic or Pacific oceans. Inuit, on the other hand, have always conceptualized the land–sea interface as continuous, and stewardship commensurate with that continuity. Ice features are named in similar ways to land features (Aporta, 2009; Aporta and Higgs, 2005), recurring seasonally. The Inuit view would have profound implications for both international waters and sector principle views of the north. Due to the Inuit presence in (what is now conceptualized as) the Canadian north, the Northwest Passage has always been a heavily contested region of the world, with the various incursions, explorations and discoveries viewed by the Inuit as invasions of their life-world by outsiders (Steckley, 2009; Eber, 2008).

The use of new mapping technologies in the north has always shaped its representation among various actors and interests. GPS and satellite-enabled visualization are pushing the vision to radical new limits that are based on but also exceed a so-called 'foundational' view of the earth as a mapped entity. The view is now extremely top-down, as satellite technology 'from above' builds upon centuries of achievement in geodesy, geomatics, surveying, and engineering measurements of the shape, volume, density, and size of the earth (Danson, 2006). The shape of the 'name' Earth (i.e. as proper name) is constantly shifting and being redefined in light of technologically induced discoveries about the earth (i.e. as noun phrase or description), as demonstrated by the recent shifting of the Greenwich mean time standard meridian approximately a hundred meters to the east of its mapped location (the standard line for well over a hundred years). The Ordnance Survey of Britain actually uses an even earlier line, an artifact of the Ordnance Survey's founding in the early 1800s (Green, 2015, page 7).

Thus, we can take nothing for granted with geographical names. They are socially constructed, defined through practices, and constantly shifting, despite continuities that can be mapped over centuries in many cases. We turn next to examining more examples of so-called post-foundational names and memes, including some found on social media through tags, some that have evolved out of older traditions, and many that have simply come to be represented in new ways, in keeping with evolving senses of geographical names, their continuities, and continued use as tools for tracking names in newly evolving NTNs.

References

Adey, Peter. 2010. *Aerial Life: Spaces, Mobilities, Affects*. Malden: John Wiley & Sons.

Afable, Patricia O. and Beeler, Madison S. 1997. Place-Names, in Goddard, I. (ed.). *Handbook of North American Indians*. Vol. 17, *Languages*. 185–189. Washington: Smithsonian Institution.

Agamben, Georgio. 1998. *Homo Sacer: Sovereign Power and Bare Life*. Stanford: Stanford University Press.

Agnew, John. 2003. *Geopolitics: Re-visioning World Politics*. London: Routledge.

Akerman, James R. (ed.). 2008. *The Imperial Map: Cartography and the Mastery of Empire*. Chicago: University of Chicago Press.

Alia, Valerie. 2007. *Names & Nunavut: Culture and Identity in Arctic Canada*. Oxford and New York: Berghahn.

Aporta, Claudio. 2009. The Trail as Home: Inuit and Their Pan-Arctic Network of Routes. *Human Ecology*. 37(2). 131–146.

Aporta, Claudio and Higgs, Eric. 2005. Satellite Culture: Global Positioning Systems, Inuit Wayfinding and the Need for a New Account of Technology. *Current Anthropology*. 46(5). 729–753.

Arlinghaus, Sandra L. and Kerski, Joseph J. 2014. *Spatial Mathematics: Theory and Practice Through Mapping*. Boca Raton: CRC Press.

Asch, Michael (ed.). 1997. *Aboriginal and Treaty Rights in Canada: Essays on Law, Equity, and Respect for Difference*. Vancouver: UBC Press.

Auge, Marc. 2009. *Non-Places: An Introduction to Supermodernity*. London and New York: Verso.

Baldwin, Andrew, Cameron, Laura and Kobayashi, Audrey (eds). 2011. *Rethinking the Great White North: Race, Nature, and Historical Geographies of Whiteness in Canada*. Vancouver: UBC Press.

Belanger, Yale D. and Lackenbauer, P. Whitney (eds). 2014. *Blockades or Breakthroughs? Aboriginal Peoples Confront the Canadian State*. Montreal and Kingston: McGill-Queen's University Press.

Bennett, John and Rowley, Susan. 2004. *Uqalurait: An Oral History of Nunavut*. Montreal and Kingston: McGill-Queen's University Press.

Berggren, J. Lennart and Jones, Alexander. 2000. *Ptolemy's Geography: An Annotated Translation of the Theoretical Chapters*. Princeton: Princeton University Press.

Braun, Bruce. 2000. Producing Vertical Territory: Geology and Governmentality in Late Victorian Canada. *Cultural Geographies*. 7(1). 7–46.

Brealey, Ken. 1995. Mapping Them Out: Euro-Canadian Cartography and the Appropriation of the Nuxalt and Ts'ilqot'in First Nations' Territories. *The Canadian Geographer*. 39(2). 140–156.

Brody, Hugh. 1975. *The People's Land: Inuit, Whites, and the Eastern Arctic*. Vancouver and Toronto: Douglas & McIntyre.

Brody, Hugh. 1981. *Maps and Dreams: Indians and the British Columbia Frontier*. Vancouver and Toronto: Douglas & McIntyre.

Bryan, Joe and Wood, Denis. 2015. *Weaponizing Maps: Indigenous Peoples and Counter-Insurgency in the Americas*. New York: Guilford Press.

Cameron, Donald and Jones, Ian G. 1983. John Snow, the Broad Street Pump and Modern Epidemiology. *International Journal of Epidemiology*. 12(4). 393–396.

Chandler, M. and Lalonde, C. 2009. Cultural Continuity as a Moderator of Suicide Risk among Canada's First Nations, in Kirmayer, L. and Valaskakis, G. (eds). *Healing Traditions: The Mental Health of Aboriginal Peoples in Canada*. Vancouver: UBC Press.

Chrisman, Nicholas. 2006. *Charting the Unknown: How Computer Mapping at Harvard Became GIS*. Redlands: ESRI Press.

Crane, Nicholas. 2002. *Mercator: The Man Who Mapped the Planet*. London: Weidenfield & Nicholson.

Danson, Edwin. 2006. *Weighing the World: The Quest to Measure the Earth*. Oxford: Oxford University Press.

Dodds, Klaus. 2014. *Geopolitics: A Very Short Introduction*. Oxford: Oxford University Press.

Donaldson, Jeffery. 2014. *Missing Link: The Evolution of Metaphor and the Metaphor of Evolution*. Montreal and Kingston: McGill-Queen's University Press.

Dorais, Louis-Jacques. 1997. *Quaqtaq: Modernity and Identity in an Inuit Community*. Montreal and Kingston: McGill-Queen's University Press.

Dorais, Louis-Jacques. 2010. *The Language of the Inuit: Syntax, Semantics, and Society in the Arctic*. Montreal and Kingston: McGill-Queen's University Press.

Dunlop, Catherine Tatiana. 2015. *Cartophilia: Maps and the Search for Identity in the French-German Borderland*. Chicago: University of Chicago Press.

Eades, Gwilym. 2012. Cree Ethnogeography. *Human Geography*. 5(3). 15–31.

Eades, Gwilym. 2015. *Maps and Memes: Redrawing Culture, Place, and Identity in Indigenous Communities*. Montreal and Kingston: McGill-Queen's University Press.

Eades, Gwilym and Zheng, Yingqin. 2014. Counter-Mapping as Assemblage: Reconfiguring Indigeneity, in Doolin, Bill, Lamprou, Eleni, Mitev, Nathalie and McLeod, Laure (eds). *Information Systems and Global Assemblages: (Re)configuring Actors, Artifacts, Organisations*. Heidelberg: Springer.

Eber, Dorothy Harley. 2008. *Encounters on the Passage: Inuit Meet the Explorers*. Toronto: University of Toronto Press.

Edney, Matthew H. 1997. *Mapping an Empire: The Geographical Construction of British India, 1765–1843*. Chicago: University of Chicago Press.

Gagnon, Alain G. and Rocher, Guy. 2002. *Reflections on the James Bay and Northern Quebec Agreement/ Regard Sur la Convention de la Baie-James et du Nord Quebecois.* Montreal: Les Editions Quebec Amerique.

Gelling, Margaret. 1978. *Signposts to the Past: Place-Names and the History of England.* London: J.M. Dent & Son.

Gelling, Margaret and Cole, Ann. 2000. *The Landscape of Place-Names.* Donington: Shaun Tyas.

Godlewska, Anne. 1995. Map, Text and Image. The Mentality of Enlightened Conquerors: A New Look at the Description de l'Egypte. *Transactions of the Institute of British Geographers.* 20(1). 5–28.

Gould, Glenn. 1967. The Idea of North. CBC Radio broadcast, December 28. www.cbc.ca/player/Radio/More+Shows/Glenn+Gould+-+The+CBC+Legacy/Audio/1960s/ID/2110447480/ (accessed August 13, 2015).

Green, Chris. 2015. There's No Need to Adjust Your Watch, but Greenwich Meridian Is on the Move. *The Independent.* August 13. 7.

Gregory, Derek. 2004. *The Colonial Present: Afghanistan, Palestine, Iraq.* Malden: Blackwell.

Hamlin, Christopher. 2009. *Cholera: The Biography.* Oxford: Oxford University Press.

Hanna, Patricia and Harrison, Bernard. 2004. *Word & World: Practice and the Foundations of Language.* Cambridge: Cambridge University Press.

Harley, J. Brian. 1989. Deconstructing the Map. *Cartographica.* 26(2). 1–20.

Harris, Cole. 2002. *Making Native Space: Colonialism, Resistance, and Reserves in British Columbia.* Vancouver: UBC Press.

Hempel, Sandra. 2006. *The Medical Detective: John Snow, Cholera and the Mystery of the Broad Street Pump.* London: Granta.

Husain, Aiyaz. 2014. *Mapping the End of Empire: American and British Strategic Visions in the Postwar World.* Cambridge, MA and London: Harvard University Press.

Isachenkov, Vladimir. 2015. Russian Submits New Claim to Swathe of Arctic Rich in Oil and Gas. *The Independent.* 5 August.

Johnson, Steven. 2006. *The Ghost Map: A Street, A City, An Epidemic and the Hidden Power of Urban Networks.* London: Penguin.

Kadmon, Naftali. 2000. *Toponymy: The Lore, Laws, and Language of Geographical Names.* Pennsylvania: Vantage.

Koch, Thomas. 2011. *Disease Maps: Epidemics on the Ground.* Chicago: University of Chicago Press.

Koch, Thomas and Denike, Kenneth. 2006. Rethinking John Snow's South London Study: A Bayesian Evaluation and Recalculation. *Social Science & Medicine.* 63(1). 271–283.

Kripke, Saul. 2011. *Philosophical Troubles: Collected Papers Volume I.* Oxford: Oxford University Press.

Kripke, Saul. 2013. *Reference and Existence.* Oxford: Oxford University Press.

Lackenbauer, P. Whitney. 2013. *The Canadian Rangers: A Living History.* Vancouver: UBC Press.

Latour, Bruno. 2005. *Reassembling the Social: An Introduction to Actor-Network Theory.* Oxford: Oxford University Press.

Laugrand, Frederic B. and Oosten, Jarich G. 2010. *Inuit Shamanism and Christianity: Transitions and Transformations in the Twentieth Century.* Montreal and Kingston: McGill-Queen's University Press.

Lefebvre, Henri. 1991. *The Production of Space.* Malden: Blackwell.

Lewis, G. Malcolm (ed.). 1998. *Cartographic Encounters: Perspectives on Native American Mapmaking and Map Use.* Chicago: University of Chicago Press.

McLeod, Kari S. 2000. Our Sense of Snow: The Myth of John Snow in Medical Geography. *Social Science & Medicine.* 50(7). 923–935.

Merrill, Robert T. 2010. *Our Magnetic Earth: The Science of Geomagnetism.* Chicago: University of Chicago Press.

Monmonier, Mark. 2015. *The History of Cartography.* Vol. 6, *Cartography in the Twentieth Century.* Chicago: University of Chicago Press.

Monmonier, Mark. 2006. *From Squaw Tit to Whorehouse Meadow: How Maps Name, Claim, and Inflame.* Chicago: University of Chicago Press.

Moore, A.W. (ed.). 1993. *Meaning and Reference.* Oxford: Oxford University Press.

Nagel, Thomas. 1986. *The View From Nowhere.* Oxford: Oxford University Press.

Niezen, Ronald. 2009. Suicide as a Way of Belonging: Causes and Consequences of Cluster Suicides in Aboriginal Communities, in Kirmayer, Laurence J. and Valasakis, Gail Guthrie. *Healing Traditions: The Mental Health of Aboriginal Peoples in Canada.* Vancouver: UBC Press.

Niezen, Ronald. 2013. *Truth & Indignation: Canada's Truth and Reconciliation Commission on Indian Residential Schools.* Toronto: University of Toronto Press.

North, F.J. 1935. *The Map of Wales [Before 1600 A.D.].* Cardiff: The National Museum of Wales.

O'Tuathail, Gerard and Dalby, Simon (eds). 1998. *Rethinking Geopolitics.* London: Routledge.

Padel, O.J. and Parsons, David N. 2008. *A Commodity of Good Names: Essays in Honour of Margaret Gelling.* Donington: Shaun Tyas.

Padron, Ricardo. 2004. *The Spacious Word: Cartography, Literature, and Empire in Early Modern Spain.* Chicago: University of Chicago Press.

Peluso, Nancy. 1995. Whose Woods Are These? Counter-Mapping Forest Territories in Kalimantan, Indonesia. *Antipode.* 27(4). 383–406.

Peterson, Michael. 2014. *Mapping in the Cloud.* New York: Guilford.

Plato. 2008. *Timaeus and Critias.* Oxford: Oxford University Press.

Ray, Arthur J. 2012. *Telling it to the Judge: Taking Native History to Court.* Montreal and Kingston: McGill-Queen's University Press.

Said, Edward. 1993. *Culture and Imperialism.* London: Vintage.

Schuurman, Nadine. 2004. *GIS: A Short Introduction.* Malden: Blackwell.

Self, Will and Steadman, Ralph. 2007. *Psychogeography.* London: Bloomsbury.

Shiode, Shino. 2012. Revisiting John Snow's Map: Network-Based Spatial Demarcation of Cholera Area. *International Journal of Geographical Information Science.* 26(1). 133–150.

Smith, Charles E. 1982. The Broad Street Pump Revisited. *International Journal of Epidemiology.* 11(2). 99–100.

Smith, George Davey. 2002. Commentary: Behind the Broad Street Pump: Aetiology, Epidemiology and Prevention of Cholera in Mid-19th Century Britain. *International Journal of Epidemiology.* 31(5). 920–932.

Smith, Gordon W. 2014. *A Historical and Legal Study of Sovereignty in the Canadian North* (ed. P. Whitney Lackenbauer). Calgary: University of Calgary Press.

Snow, John. 1855. *On the Mode of Communication of Cholera.* London: John Churchill.

Sparke, Matthew. 2005. *In the Space of Theory: Postfoundational Geographies of the Nation-State.* Minneapolis: University of Minnesota Press.

Steckley, John L. 2009. *White Lies About the Inuit*. Toronto: University of Toronto Press.

Steinberg, Philip E. 2001. *The Social Construction of the Ocean*. Cambridge: Cambridge University Press.

Strabo. 1917. *Geography*. Cambridge, MA: Harvard University Press.

Tomaszewski, Brian. 2015. *Geographic Information Systems (GIS) for Disaster Management*. Boca Raton: CRC Press.

Vandenbroucke, J.P., Eelkman Rooda, H.M. and Beukers, H. 1991. Who Made John Snow a Hero? *American Journal of Epidemiology*. 133(10). 967–973.

Vinten-Johansen, Peter, Brody, Howard, Paneth, Nigel, Rachman, Stephen and Rip, Michael. 2003. *Cholera, Chloroform, and the Science of Medicine: A Life of John Snow*. Oxford: Oxford University Press.

Wood, Denis. 2010. *Rethinking the Power of Maps*. New York: Guilford.

Wright, Shelley. 2014. *Our Ice Is Vanishing/Sikuvut Nunguliqtuq: A History of Inuit, Newcomers, and Climate Change*. Montreal and Kingston: McGill-Queen's University Press.

6 Neogeographies of the name

Naming the Anthropocene

The earth is constantly evolving 'new' geographies through shifting tectonic plates, extinctions, speciation, and the adaptations of human culture. Catastrophic changes occur at various scales in the form of, for example, meteor impacts, political revolutions, and human-induced climate change. The term Anthropocene offers a case in point—the name of a 'thing' that is both political and contested while maintaining the aura of an objectively 'measured' entity in the world, a time frame that puts humans in their place on a geological timescale. But what is this thing we call the Anthropocene? It is, first and foremost, a marker for indicating "a fundamental change in the relationship between humans and the Earth system" (Lewis and Maslin, 2015, page 171).

The 'baptism' of an epoch of earth-time defined by human influence can be traced to a theologist, Thomas Jenkyn, and a reverend, Samuel Haughton, who together called the present time the Anthropozoic, i.e. a time when the very geology of the earth can in its essentials be defined by the presence of a human signature in its strata (Lewis and Maslin, 2015, page 172). At present there is no agreed-upon start date for the Anthropocene, but it is accepted as a human-created term for referring to an epoch that both exceeds humanity itself and is defined by that humanity, for better or worse. The very idea of named divisions in earth-time indicates a very human need to name temporal phenomena expressed across geographic space and in the depths of the earth's strata. The idea of the Anthropocene is therefore both reflexive and subjective, when the subject is defined as the human species itself. Other species do not have this notion of the Anthropocene nor, presumably, did proto-human species. Would a future race of post-humans define things in the same way? What about a Martian species? These are matters of pure speculation, but they are serious questions to be considered by philosophers and geologists/geographers alike.

For the present, should the Anthropocene, as a name, become established, it will be a particularly acute tool for political change. The start date itself will cast judgment on countries associated with, for example, the start of the industrial revolution; on the other hand it could end up playing into the hands of climate change deniers should the date be set before the impact of modern-day humanity (Lewis and Maslin, 2015, page 171). We have come to associate humanity with catastrophic change despite that position being often associated with radical left sentiment (Huggett, 1997). Uniformitarianism is seen as a reactionary position in some circles, for if the past is key to the present, perhaps not much has changed at all and we are simply framing things to suit our subjective biases and inclinations.

Take the idea of names in Thomas Hardy and how this has come to clash with, even counter-map, ideas of how to counter climate change and global warming in parts of England associated with what Hardy called 'Wessex': Devon and Dorset counties. Hardy's novels usually contain a map of Wessex, with fictionalised versions of real names corresponding one-to-one with the real world. We consider here one of Hardy's (1978) most famous novels, *Tess of the D'Urbervilles* (hereafter *Tess*), juxtaposing it against a contemporary push to place wind farms on the coasts made famous by Hardy. The following is reproduced from text accompanying a map placed at the beginning of *Tess* (page 25):

> Hardy's Wessex is so familiar that it is hard to realize how odd it is that a novelist should have tied himself by so many strings to a particular tract of territory. Many novelists have set their scenes in real places, or have written with some features of a familiar landscape always before them. But Hardy has done something different. Almost every step taken by his characters is taken along real roads or over real heaths; the towns and villages, the hills, even many of the houses, are identifiable. It is as if Hardy's imagination could not work unless with solid ground under its feet, with solid objects to be seen around it. Many of the characters, there is little doubt, contain more or less of one real person, more or less of another, with elements drawn purely from imagination or from the accumulated layers of experience, which comes to much the same thing. But with the topography, Hardy was rarely satisfied with anything less than a one-to-one correspondence between the fictional and the real.

Furthermore, Hardy's *Tess* is intimately caught up in the question of names, both personal and geographical. Tess, in the text, most often carries a degraded form of her last name, Durbyfield, indicating that the time of any 'greatness' of her lineage is long past. Tess's attempt to reconnect with that past is a large part of the origin of the tragedy that ensues, as she is taken advantage of by a usurper of the older form of the name, a rich landowner named Alec.

Near the beginning of *Tess* we are introduced to her by way of a group of women performing a ritual in some ways similar to beating of the bounds, accompanied by dance (i.e. a harvest celebration). Tess is introduced as an example of womankind in the context of the landscape of Wessex and of England as a whole. The boundaries of Tess's world are described through names and in a language that expresses a feeling for the material at hand:

> This fertile and sheltered tract of country, in which the fields are never brown and the springs never dry, is bounded on the south by the bold chalk ridge that embraces the prominences of Hambledon Hill, Bulbarrow, Nettlecombe-Tout, Dogbury, High Stoy, and Bubb Down. The traveller from the coast, who, after plodding northward for a score of miles over calcareous downs and corn-lands, suddenly reaches the verge of one of these escarpments, is surprised and delighted to behold, extended like a map beneath him, a country differing absolutely from that which he has passed through.
>
> (Hardy, 1978, page 48)

The boundaries of the named world are differentiated and distinct for Hardy's characters travelling across it on foot. Tess's character is formed as much by the landscape as by her name, and both are evolving rapidly beyond their original foundations in a changing and industrializing world. Hardy often notes how large sections of Wessex have become homogenized by the railways but that distinct sections, such as that described above, remain. Depending on where one draws the boundary line (Lewis and Maslin, 2015, page 171 posit either 1610 or 1964 as the beginning of the Anthropocene), one can posit that this change is part and parcel of what it means to be human in that most human of epochs, the Anthropocene. The irony, though, is that many of the rituals and routines of humankind are changing beyond all recognition:

> The forests have departed, but some old customs of their shades remain. Many, however, linger only in a metamorphosed or disguised form. The May-Day dance, for instance, was to be discerned on the afternoon under notice, in the guise of the club revel, or 'club-walking', as it was there called.
>
> (Hardy, 1978, page 49)

This procession is unusual in that it consists, in this instance, of a group of women including Tess, celebrating Ceres the goddess of agriculture, described in terms similar to rogation:

> In addition to the distinction of a white frock, every woman and girl carried in her right hand a peeled willow wand, and in her left a bunch of white flowers. The peeling of the former, and the selection of the latter, had been an operation of personal care.
>
> (Hardy, 1978, page 50)

One can see how, through Hardy's evocation of the dying rituals of the past and his naming of the landscape, one of its exemplars, Tess of the ancient D'Uberville family, is similarly dying through the degradation of her name and the denudation of her body through the acts of men like Alec and her second love Angel Clare. This is equally a story of love of landscape and the geographical names of which it is composed, which even today refer to the maps spread out before us as we traverse its breadth and depth by train or car. This is the staying power of Hardy, and the reason why his work is so prescient for beginning to unpick some of the complexities of and counter-mappings enabled by the term 'Anthropocene.' The new geographies of Hardy's Wessex are fraught with tension introduced by Hardy's language of landscape and by the way he described devastating changes to both that landscape and the people whose livelihoods depended upon it on a daily basis.

For the purposes of the present discussion we note that a recent proposal to install wind farms in response to the future inability of humans to fully capitalize on fossil fuels (because of climate change) has met with resistance from current residents of Hardy's historic Wessex (*Guardian*, 2014; *MailOnline*, 2013). The people of Dorset, upon which Hardy modeled Wessex, oppose this human response to climate-induced change, i.e. the installation of wind farms along their coast. This is the very coast described as having been traversed by Hardy's characters, who can in turn be posited in some cases as the ancestors of Dorset's present-day inhabitants. The Wessex of Hardy has become a meme for counteracting change in Dorset, change represented by both the climate and developments brought in, like the railways and the industrial revolution, from outside (Eades, 2015c). Many issues of change are the same as in Hardy's day, and it is through naming of the landscape and mapping of that same landscape through its names for the production of memes that protection is enacted. Place-based practices link up with language to track names in NTNs through time, counteracting the language and practices of those in favor of the wind farms. While Hardy's landscapes seem timeless, the technological practices of which they are composed are changing rapidly in keeping with what Hardy elucidated so clearly, and this is the ultimate irony. New tools for making maps of 'Wessex'/Dorset are available and it is to these that we now turn.

Counter-mapping in the cloud

New geographies of the internet age are referred to as 'neogeographies' and therefore have a very particular scale of temporal reference, one that is even more recent than a term like 'Anthropocene'. Within the discipline of geography, a great deal of debate has revolved around the idea of *neo*geography, often posited as being pitted against *paleo*geography, or old ways of doing things, prior to internet mapping, mashups, and the like (Leszczynski, 2014). A map mashup is an online map, usually one that shows geographical names, made using a proprietary base overlain with a 'custom' dataset (as noted in the previous chapter). An example of a proprietary base is Google Maps. An example of a custom dataset would be the place names of Hardy's Wessex plotted as exact locations, with labels corresponding to point, line, or area features in the landscape. Google Maps default base map shows the city of Exeter, but the Hardy/Wessex map would also label the town Exonbury, the fictional correlate in Hardy's novels, at the western edge of his Wessex. The point of such a map would be to add another layer of meaning to an already existing 'commonsense' map of the area Hardy fictionalized and, in doing so, to counter-map what we think we know about that area. Adding all of Hardy's place names to a Google Map (i.e. creating a Hardy mashup) would in effect create a counter-map useful for the purposes of counteracting a hegemonic or powerful message from outside. That message might be, for example, that Dorset needs wind farms.

As noted above, counter-maps (in the cloud, i.e. on Google or some other proprietary base) can contain multiple (more than two) representations of the same places or areas of territory: for example, photographs of historic buildings found in Hardy's novels could be added to Google Earth in order to deepen the counter-map's message and more powerfully assert its counter-hegemonic message. Google Earth uses the keyhole markup language (kml) file type for the production and storage of geographic primitives (points, lines, and areas) and associated information. It also facilitates the use of .gpx files, or those produced using GPS, often used by psycho-geographers and other neogeo-graphers and counter-mappers to map and upload their maps and data to the cloud. The related ideas of data types, platforms, and translation of file types is central to new assemblage-based methods for the production of cloud-based maps and counter-maps (Eades and Zheng, 2014; Peterson, 2014).

The idea of the 'cloud' itself needs more elucidation (Hu, 2015). Computing is increasingly moving away from reliance on desktop-based memory and processing towards the use of offsite (distant) servers for storage and com-puting tasks (services). With processing done away from the platform of the user, the latter becomes less important (or more flexible) in terms of speed and power but more so in terms of how the map is visualized. Mobile devices such as phones, tablets, and laptops devote a good deal of memory and hardware to visual processing, allowing for high-quality maps, images, and video storage and use. Many of these same devices are now also equipped with a GPS that can be used in conjunction with an app (application) to make maps of, for example, paths walked, exercise routes, or tourist itineraries. When uploaded into the cloud, the .gpx or .kml file is copied to an external server and can be activated so that it is shared across the network to other registered users (for example on Google Earth Community) for visualization or modification (depending on the configuration of the site). A site like Map My Run or Map My Ride lets users share their exercise routes, which others can comment on (but not modify).

The idea of counter-mapping as an assemblage of data transforms the counter-map into a much more flexible, but also more diffuse, concept. It opens up the idea that, by simply moving through a space and creating a trace of that movement, a counter-map can be an almost unconscious creation. Mobile-phone users often leave traces, or metadata, without realizing they are doing so (Pomerantz, 2015; Shifman, 2014). Metadata is collected by advertisers and government agencies, often without the knowledge (or consent) of the users they target. This metadata frequently includes locational information, easily turned into maps by anyone possessing the metadata in the form of sets of coordinates showing the where and when of a user's location, of particular interest to state surveillance bodies and/or corporate offices interested in tracking consumer behavior. Surveillance literally means 'map survey' and the idea has been turned on its head by a sub-culture of mappers interested in what they call the mapping of everyday life, or sous-veillance (mapping from below). This kind of counter-mapping can also take the form of mapping the mappers, by producing maps of least surveillance, which are paths that avoid CCTV, for example, or maps that point out the locations of such cameras (though it is illegal to map CCTV in the UK) (O'Rourke, 2013).

Many artists use locational data or the 'locative' (to use William Gibson's terminology) to structure observations about space and to comment on or critique societal aspects of space, state spying activities, and the like through the production of locative art (Jones, 2013; Gibson, 2007). Often helped by the observations or comments from artists, the whole idea of counter-mapping is being turned on its head due to the association of mapping with surveillance. This has been brought on, I argue, by technologies such as Google Street View, which produce ground-level views of cities, including their residential areas.

Battle to blur

Some residents of cities are fighting to (re)blank space, in essence, counter-mapping Google's Street View by blurring the space in which their homes exist on that view. The reasons cited are many including, foremost, maintaining the right to privacy. This represents a very fine-grained and targeted form of counter-mapping with strong cultural roots. For example, the push to 'blur' houses on Street View has been strongest in Germany, where whole communities have banded together to blur out their living spaces. There are other examples of this new push to 'blur' maps, and we posit these new counter-mapping impulses as part and parcel of neogeographical mapping sensibilities and the democratization of mapmaking. When anyone can zoom in to see your house you suddenly become conscious of privacy issues, especially when, as in some cases, the Google Street View cameras have made it possible to see inside houses and have caught people out in nefarious or anti-social activities, such as adultery or drug use. This is apart from everyday notions of privacy (Nagel, 2002) that should allow residents baseline confidence that their lives can be lived without being recorded or viewed by anonymous, corporate, or governmental actors (not to mention neighbours or other known individuals). The reasons behind Germany's stronger push to assert privacy in the face of Street View are not difficult to surmise, given the legacy of state interference (i.e. the Stasi) in the lives of its citizens during the Cold War era.

The impulse to blank Street View maps is opposite to that which drives indigenous counter-mapping efforts in, for example, Canada, where things have progressed to the point that Inuit mappers working for Google create

Street View maps by bicycle and snowmobile (Austin, 2012). The converse, new, impulse to blur maps can be attributed to new scales of mapping whereby society's collective map has attained a degree of detail seen as counter-productive. The Borges and Carroll maps come to mind here. In Borges' (2000, page 139) one-paragraph short story "On Scientific Rigor," this idea is explored with exemplary brevity:

> In the Empire in question, the Cartographer's Art reached such a degree of Perfection that the map of a single Province took up an entire City, and the map of the Empire covered an entire Province. After a while these Outsized Maps were no longer sufficient, and the Schools of Cartography created a Map of the Empire that was the size of the Empire, matching it point by point. Later Generations, which were less Devoted to the Study of Cartography, found this Map Irrelevant, and with more than a little Irreverence left it exposed to the Inclemencies of the Sun and Winter. In the Western desert there are scattered Ruins of the Map, inhabited by Animals and Beggars. No other relics of the Geographic Discipline can be found anywhere else in the Land.

The question here is, has Google achieved what the 'Empire' could not, and if it has, what parallels exist with Borges' parable? In light of what is happening in neogeographical mapping vis-à-vis Google, it is fruitful to examine his parable more closely, deconstructing it. First, we have the correct association of the cartographer's art with empire. Cartography was invented to further state interests, and the interests of (European) states in the Enlightenment meant 'spreading the word' through colonialism and the extension of imperial power. This was achieved in large part through maps, as has been argued by many scholars of critical cartography (Husain, 2014; Akerman, 2008; Edney, 1997). The art (and science) of cartography was indeed all-encompassing and perfectionist in impulse, in the sense that cartographers sought to impose order and control on space. That space, in turn, held real (indigenous) bodies subject to colonial rule from afar through the performance and inscription of the cartographer's map, produced in the name of the sovereign (imperial ruler).

Borges introduces the idea that maps can become 'outsized', which is of course absurd. But he makes a good point: the cartographer's (and empire's) impulse is all-encompassing and also absurd in the end. Maps tend to freeze spatial knowledge, but the objects, people, and relationships on the ground continue to 'move', to change and evolve. One could indeed imagine a very large map produced out of larger and larger-scale map series produced for whole nations and empires that, if joined at the edges, could very well cover a good deal of territory. But Borges' absurdity is not literal,—his metaphor operates to question the mapping impulse (and by extension, science) itself. The empire of Borges' poem goes so far as to cover itself with a map. The intergenerational aspect comes in—younger people do not respect the map in the same way and it is allowed to degrade. The parallel here (in the real world, not in Borges' story) is the recent decline in cartographic knowledge that society is seeing with the advent of neogeography.

Neogeographers are, in essence, coders who do not need older (read paleogeographical) knowledge of the cartographer's craft in order to make their maps. New maps are produced 'on the fly', as needed, in response to immediate needs, for example on a mobile device such as a phone (Turner, 2006). The parallel with Borges' story is that older (paper) maps are being allowed to 'lapse', degrade, and erode into nothing—the paper literally tears, rots, and blows away in the wind. The old maps are now the refuge of animals and the homeless because they have a physical presence. Newer virtual maps in cyberspace, on the internet, and on devices, represent the Google generation's way of interacting with space. The neogeographical generation is thoroughly comfortable with 'file types' such as .kml or .gpx and not averse to what is called 'code', for example markup languages, of which .kml and html (hypertext markup language) are just two of many types (the generic of which is xml, or extensible markup language). These can also be seen as kinds of 'metadata' for creating 'content' on the web in way that is structured but not limited by style of presentation (which is a separate issue). Geographic markup languages (gml) bring cartographic concepts such as projections, coordinate systems, and locations into the fold and are used in a variety of new mapping applications online, including the Atlas of Canada (Natural Resources Canada, 2015).

But doesn't Google's one-to-one online map of the world run the risk of alienating the next generation? Who will take refuge under the rotting hunks of Google's maps once society abandons the new corporate impulse to map every last space on earth in order to, in essence, colonize the earth with its generic-looking, non-cartographic (i.e. neutrally styled) maps? The main question we need to ask here is: how do new mapping sensibilities, platforms, and data types affect society's ability to track geographical names through time? What kind of new memes for the transmission of those names are enabled by neogeography? Does the radical break with the past implicit in neogeography mean that knowledge will be lost? As seen in the previous chapter, a large part of the counter-mapping impulse is driven by preservation and evolution of indigenous and local geographical naming practices. I do not see any reason why the advent (and advance) of neogeography necessitates the loss of those practices, nor of the knowledge they are designed to pass on and preserve. Change has always been part of geographical naming, with old names changing, morphing, or evolving while retaining traces of their older selves, as philologists have amply demonstrated (Gelling and Cole, 2000). At the same time, indigenous groups such as the Cree, the Inuit, and the Tlingit are constantly adding new geographical names to their knowledge storehouses, as seen, for example, in the production of a Tlingit names atlas for the southern coast of Alaska (Thornton, 2012).

Indigenous groups themselves are increasingly going online with the creation of mapping platforms for preserving and transmitting traditional land-based knowledge systems (Cope and Elwood, 2009). There is a trade-off with such efforts, however: like previous cartographic productions they tend to 'freeze' knowledge in a snapshot in time unless there are repeated surveys and frequent updates to the geographical names database. This static sense of mapping (which is not solved in GIS or neogeographical systems) is compensated for by the fact that the knowledge has been (counter-)mapped in the first place, reifying knowledge that was previously held in bodies, brains, and oral retellings of 'spatial stories' (Eades, 2015b; Johnson, 2010; Johnson and Hunn, 2010; Collignon, 2006; Stern and Stevenson, 2006). In essence, geographical NTNs are evolving from embodied and performed formats to more inscriptive cartographic regimes. Both are 'virtual' in Deleuze's sense of remaining as traces beneath the surface of the map until activated through the necessities of use (Deleuze, 1990; Bergson, 1990).

More mainstream sites representing potential neogeographical NTNs (though not designed explicitly as such) include Geonames and Wikimapia. The georeferencing of names into online gazetteers faces technical challenges greater than with 'traditional' gazetteer production, which may or may not include a map component. The sheer number and variety of map-viewing platforms for interacting with geographical names, often while mobile, is staggering, and only set to increase. With the use of on-board GPS-enabled navigation systems for directing drivers there are also renewed ethical and safety aspects, though these have been present in geographical naming practices from time immemorial (Hill, 2006). A major component of Cree geographical names and their accurate communication and transmission has to do with safety (Wellen, 2008). In contemporary society, too, safety is very much dependent on accurate and up-to-date gazetteers, correctly georeferenced and mapped, especially in terms of 'ontology' or the geographical nature of the named feature. This can include part-whole (mereological) relationships (Evans, 1982), as well as the idea of geographic primitives (i.e. vector models in GIS made up of point, line, area, and volumetric features). With mobile devices, new vernacular geographical names are being added to 'official' gazetteers at an accelerating rate, with emergency services scrambling to keep up (Lawrence, personal communication).

The 'battle to blur', mentioned above as a kind of counter-mapping to official geographical naming systems that attempt to erase or blur features from the landscape in the name of privacy, is also being activated in the name of more nefarious, larger-scale processes, including international terrorism. As explored below, the Islamic State of Iraq and Syria (ISIS) is a neogeographical name for a group that has made powerful use of ideas of counter-power (and -mapping) to make major gains in the Middle East. Part of their power lies precisely in spatial depictions of their ever-expanding territory and in the usurpation (and often destruction) of named places associated with areas of great archaeological and cultural significance (Cockburn, 2015).

(Geo)spatial interpellation, nationalism, and ISIS

A new kind of counter-mapping is taking place on the ground in Syria, Iraq, and Turkey, with the most interesting 'subject position' being that of the

Kurds, a stateless group inhabiting parts of all three countries (Sengupta, 2015, page 31; Adler and Srebro, 2015, page 167). The driver for the current high profile of the Kurds (on the world scene, with Kurds often appearing in stories in British newspapers) can in part be attributed to the rise of the radical Islamist group Al Qaeda in Iraq (Cockburn, 2015, page 73). The influence of this extremist Sunni Muslim group spread to Syria, becoming the Islamic State of Iraq and Syria (ISIS), now with ambitions to create its own 'state' or caliphate with disregard for the historic Sykes-Picot line that divided British and French interests after World War I (Adler and Srebro, 2015, page 164; Barr, 2011). ISIS, before breaking onto the world stage with a series of high-profile televised and streamed executions and destruction of ancient archaeological sites, had been part of an informal coalition of groups fighting Assad's regime post-Arab Spring. With money and weapons at its disposal, much of it captured from corrupt Iraqi soldiers and officials, it soon demonstrated itself to be 'a cut above' the rest, with the ability to collect taxes, raise a standing army, and create (i.e. enforce) loyalty within its so-called caliphate.

Earlier, Syria had been part of a 'battle to blur' its atrocities on the ground, including the razing of suburbs known to contain opposition forces, such as the "poor district of Masha al-Arb'een in Hama" shown on satellite images before and after its destruction (Cockburn, 2014, page 27). Assad was remapping cities that had fallen into enemy hands, regaining territory block by block but also seeking to hide, through sheer obliteration of the terrain, what in fact probably amounted to war crimes, including the use of chemical weapons, to aid in clearing areas of opposition. Assad's counter-mapping worked until his forces began to weaken due to the length of the war. At this point the rise of ISIS became apparent and threatened to become an even bigger threat to the west than Assad himself, causing a rapid shift of allegiances in the region. One of the most striking things throughout various conflicts in Syria and Iraq has been the production of maps by western media to show the spread of ISIS. These maps are of two main types. The first 'objectively' shows areas of current control on the ground by ISIS forces, with 'tentacles' joining various captured cities: Mosul, Tikrit, Fallujah, to name just a few of the set of key settlements captured through brute force and through the negative power of the ISIS ideology to 'interpellate' local Sunni (and Shia) into its sphere of control. The second type shows where ISIS allegedly hopes to spread, with whole countries—including Egypt, Libya, Tunisia, Syria, Iraq, and Spain—blacked out to indicate their capture by the black flag of ISIS.

Zizek (1993, pages 73–77) has explored how this kind of nationalistic (spatial) interpellation operates from the perspective of radical psychoanalysis. Zizek's method of analysis of nationalism is unorthodox but also entirely appropriate given the willingness of ISIS to embrace the most extreme and virulent forms of ideology, justifying rape, torture, and genocide, both cultural and physical, of whole swathes of the human race. Zizek posits that the subject of interpellation is a bit like a prisoner in game theoretical space calculating the position that will best maximize the outcome for him- or herself. A doubly retroactive logic is at play here: the subject is always already part of the 'big other' calling out to him or her, they just don't know it yet. After identification (with ISIS) occurs, it seems to the subject (according to this theory) that it was pre-ordained, addressing something that was there all along just waiting to be activated. This is what is meant by there being something that is 'in you more than you', an ideological kernel of self. But Zizek goes beyond Althusser's (1977) original formulation of interpellation, at the same time giving it a decisive 'spatial' (nationalistic) twist. For Zizek, this other level of inter-pellation, beyond the 'always already' aspect, is like a game in which recognition of others in a group seen as like-minded, and thus ready to commit to the ideological moment of conversion (or interpellation), is key. Where inter-pellation for Althusser was individualized, it is collective for Zizek, and thus all the more effective for being able to 'convert' more individuals in a sweep.

This is precisely what ISIS does, with its flags, Twitter and social-media pronouncements and postings, and with a good deal of help from the west. I posit here that the map of ISIS is a key component of the group's success and that it underlies a good deal of its ideological appeal, one that literally blurs boundaries, re-uniting territories that were for decades artificially divided (between Iraq and Syria). The first kind of map mentioned above shows ISIS like a kind of organism, or virus, spreading a web across Syria and Iraq, with cross-connections between cities in northern parts of Syria and western parts of Iraq, pushing northward against Kurdish-occupied areas, Turkey, and features resistant to its spread. The second kind of map (again produced mostly by western journalists) props up ISIS in an even worse way. It posits the future success of its spread across national boundaries, with recapture of areas such as Spain and northern Africa deemed part of a lost Islamic golden age. Fisk (2015, page 28) has commented upon this phenomenon:

> We have nightmares. They have dreams. Isis provides real or fantasy nightmares almost every day. A Croat is beheaded in Egypt. An Isis suicide bomber kills almost 70 civilians in a Baghdad market. Twitter, Fox, ABC News and the British tabloids bring us an 'Isis map' – which may be a load of old baloney for all they know, because Isis has said nothing about it. The map purports to show us just how much of the globe Isis intends to swallow at first gulp: Spain, Greece, the former Yugoslavia, Turkey, the Levant, Egypt, the Maghreb, half of Africa, all of India, Pakistan, and a chunk of China.

ISIS operates through names not only by way of spatial interpellation and maps, as explored above, but also literally through names. Obviously names are parts of maps in the sense that captured cities form part of the sensible victory of the spread of ISIS ideology. But there are other ways that names are important. Consider person names. Cockburn (2015, page 6) reports that group interpellation of even moderate Muslims is almost inevitable given the close-knit nature of most Muslim communities:

A striking development in the Islamic world in recent decades is the way in which Wahhabism is taking over mainstream Sunni Islam. In one country after another Saudi Arabia is putting up the money for the training of preachers and the building of mosques. A result of this is the spread of sectarian strife between Sunni and Shia. The latter find themselves targeted with unprecedented viciousness, from Tunisia to Indonesia. Such sectarianism is not confined to country villages outside Aleppo or in the Punjab; it is poisoning relations between the two sects in every Islamic grouping. A Muslim friend in London told me: 'Go through the address books of any Sunni or Shia in Britain and you will find very few names belonging to people outside their own community.'

Another aspect of names is what is, or is not, sanctioned by media and the state as the 'official' name of ISIS. Egypt has created a list of names to call Isis that avoid stereotyping Islam in a negative way. Thus, Pizzi (2015) notes that the government of Egypt deems that while the terms assassin, slayer, destroyer, eradicator, terrorist, extremist, criminal, savage, and murderer are all acceptable, jihadi, jihadist, Isil, Islamic State, and Islamist are all too negative or biased to gain official sanction. What to call Isis is indeed a highly contentious issue, one that is caught up in the spatial and interpellative logic of the group. Cockburn (2015, page 43) again summarizes the issue very well (commenting upon the term for ISIS that is favoured in the United States, which is ISIL):

> Its very name (the Islamic State of Iraq and the Levant) expresses its intention: it plans to build an Islamic state in Iraq and in 'al-Sham' or greater Syria. It is not planning to share power with anybody. Led since 2010 by Abu Bakr al-Baghdadi, also known as Abu Dua, it has proved itself even more violent and sectarian than the 'core' al-Qaeda, led by Ayman al-Zawahiri, who is based in Pakistan.

ISIS is like a meme that colonizes the brains of those it converts, and it achieves this in large part through the force of its names. I argue here that ISIS is building its own new and highly virulent NTN for spreading the word about the new state or caliphate, one with ambitions (according to the west) well beyond Syria and Iraq. Whether or not this is true remains to be seen, and hopefully for the sake of freedom and liberty things will not go that far. ISIS fanaticism and fascism are enabled in large part by the tools for naming at its disposal: social media, maps, ideology, flags, and the power of its army and officials to raise capital and command through force. The strength of the opposition forces hoping to stem the tide of the virulent ISIS ideology is that they have the same forces in their hands, as well as the condemnation and collaboration of multiple (and global) forces opposed to it. It is a matter of coordinating the various oppositions, including crucially the stateless Kurds, in such a way that the tide is stemmed rather than channeled or inadvertently spread.

The hubris of proactive disaster mapping

Blackboxing (Pasquale, 2015) is the algorithmic association of a name with a negative event or outcomes. *Spatial* blackboxing is the algorithmic association of a *geographical* name with negative events or outcomes (Eades, 2015a). Blackboxing is distinguished from the Zizek/Althusser senses of interpellation in that the former (blackboxing) is an unwanted, negative identity that is formed online outside of the named person's (or place's) control. The latter (interpellation) takes on a 'positive' identification in that the person (or place) named by the interpellation identified the part that was 'always there' and that makes the interpellation 'stick', for example, when a young person joins ISIS by first becoming converted online and then travelling to Syria to fight. Far from wanting the negative association to stick, however, a named individual subject to blackboxing (for example, association with a negative credit score or a past crime or misdemeanor) would often like nothing more than to see the algorithm somehow 'forget' the association. Google has become mired in legal challenges associated with the European Court's ruling dubbed the right to be forgotten (Harper and Owen, 2014, page 1). This could conceivably apply to those who have traveled to Syria to fight with ISIS but, following a subsequent de-conversion (or re-interpellation into non-ISIS, 'western' ideology) seek to return home, with actions possibly recorded in online videos, tweets, and emails forming evidence of their past behaviors.

All of this is somehow caught up in the logic of revelations associated with rogue analyst Snowden's release of information on how the governments of the United States and United Kingdom keep track of personal data and communications in cooperation with corporations like Google, Facebook, and Twitter (Bowcott and Ball, 2014; Vincent, 2014). Before turning to how spatial blackboxing can, in theory, create negative associations for places targeted by well-meaning organizations' mapping efforts, I will first look at how it affects individuals. Pasquale (2015) has explored the phenomenon in most depth from scholarly and legal perspectives. I use several of his observations as the basis for further speculations on how algorithms can be used to accumulate (positive or negative) knowledge about places. Pasquale (2015, page 72) notes that,

> Google came under fire in 2012, for example, in an awkward situation regarding a prominent German woman, Bettina Wulff. Users who typed her name into the search box were likely to see 'bettina wulff prostituierte' and 'bettina wulff escort' appear in the 'auto-complete' list underneath. Those phrases reflected unfounded rumors about Wulff, who has had to obtain more than thirty cease-and-desist orders in Germany against bloggers and journalists who mischaracterized her past salaciously. Wulff feared that users would interpret the autocompletes (which Google offers as a convenience to users) as a judgment on her character rather than as an artifact of her prolonged and victorious battles against slanderers.

At stake is not only the individual's right to privacy and freedom from unwanted and slanderous negative associations with their names. There also exists a right to freedom from government snooping, and large corporations have been demonstrated (i.e. by the Snowden revelations) to be working with organizations such as the National Security Agency (NSA) in the US and Government Communications Headquarters (GCHQ) in the UK allowing 'backdoor' access to metadata and other communications information. These two aspects, 1) algorithmic search and control (with associated long-term, and public, memory) and 2) government and corporate complicity in 'harvesting' of large datasets (also known as 'big data') and metadata (or data about data), have deep implications for individuals, corporations, and places. A moral scale of impact can be posited as heaviest on the individual, who is, after all, powerless in the face of corporate and governmental networks. Corporations—for example, those associated with sketchy or illegal practices such as insecure mortgage-backing and interest rate manipulations—can also find themselves in negative blackboxes.

Pasquale (2015, page 118) reminds us that, unlike individuals, corporation names refer differently than do individual names:

> When we refer to, say, 'Citigroup,' we have some sense of that being a *single* company. But firms that large (it has over 250,000 employees) are really composed of hundreds of entities, over which the central office has varying degrees of control. Citigroup has subsidiaries controlled via ownership of shares, the same way a majority stockowner may have voting rights over a firm. Contracts may bind the directors of one firm to always obey instructions from directors of another firm, or to remit all their earnings to a 'parent' company.

Shell companies, or companies within companies, and the production of dodgy paper trails are precisely what led to streams of foreclosures in the United States just prior to the economic crash of 2007–08. Such shell companies often had unfamiliar-sounding names, which was enough for some debtors to abandon their responsibilities and default upon (i.e. dump) mortgage payments. While a company like Google or a government entity would purport to be above such practices, they are both still corporate entities and, as such, lack the moral imperative or drive that keeps individuals 'in check.' Without the individual and with shell entities or other parts of government to whom the buck can be passed (such as in government surveillance where someone else can always be blamed), there is not an equivalent imperative set of interests to protect. This is what makes the Snowden revelations so remarkable: one individual, acting alone at first, took on myriad governmental and corporate entities and practices. The result has been directly measurable improvement of individual civil liberties in North America and Europe (though some would argue that security has been correspondingly reduced, an argument we will not address directly here).

Places are another case entirely. Unlike both individuals and corporations, places can be seen as composed of moral entities distributed in a sort of communitarian local space. But places also exist in phenomenological time as continuously existing entities with material (embodied) and mental (ideal) aspects, evolving without decisively breaking with the past (Malpas, 2006; Casey, 1993, 1997, and 2002). There is an ethics of place that is both fluid and contested, as the case of the Missing Maps project, run by Humanitarian OpenStreetMap Team (HOT, at hotosm.org) demonstrates below (Michael, 2014).

I argue here that places subject to natural disasters (often exacerbated by poor human responses to those disasters) such as New Orleans (hit by a hurricane in 2005) or Nepal (which experienced a devastating earthquake in 2015) can become blackboxed through association with negative events in the media. The idea of 'missing maps' has become a kind of meme that has colonized otherwise well-meaning neogeographers' brains. A sub-culture within neogeography (a sub-culture in itself) is concerned with proactively mapping areas predicted to be at future risk of natural disaster. Organizations such as HOT have taken it upon themselves to comprehensively map developing areas where the online volunteered geographic information (VGI) site Open-StreetMap is blank or insufficiently filled in. The concept of VGI requires some explanation. It is, essentially, the idea of turning citizens into sensors (Goodchild, 2007) for the collection and uploading of geographical informa-tion, including all sorts of geographic primitives (points, lines, and areas). Point-based information is the most popular form for upload to a site like OpenStreetMap, with line-based information (e.g. a GPS-generated line representation of a road) and areal information being next in popularity (because the latter are more technical to produce).

OpenStreetMap exists as a legacy of Ordnance Survey, which opened up a base dataset to the public for additions and modifications of VGI (Ramm et al., 2010). Like Wikipedia, anyone can change the map and anyone can change what someone else has posted. Besides being a mapping platform, Open-StreetMap is part of a sub-culture that conducts local mapping parties to fill in blank spaces on the map. As such, OpenStreetMap is, de facto, a counter-map that is participatory, public, and driven by grassroots efforts. It fulfills a long-standing dream in the GIS world that sought the 'true' public participation GIS (PPGIS) platform for including non-experts in mapping processes in that anyone can use it and it does not require a great deal of technical knowledge (Sieber, 2006).

What the Humanitarian OpenStreetMap Team (HOT) is doing, through their project Missing Maps, is leveraging the power and openness of Open-StreetMap to fill in blanks in anticipation of future disasters. This means, primarily, filling in place names and roads for accessibility and navigation. There is real need for such mapping, as was demonstrated, for example, in Africa when Ebola victims (and their families) in need of help met with delays in the delivery of medicines and equipment for burial due to the lack of a base map for navigating to the location of the victims (Economist, 2014). HOT identified OpenStreetMap as the best available tool for training locals to fill in their own maps, and started sending teams around to areas that had already experienced disaster and were expected to do so again in the future. There are three related critical issues of relevance to such efforts: 1) the issue of identification and potential blackboxing; 2) the issue of OpenStreetMap being used in a top-down and expert-driven way by outsiders (non-locals); and 3) the issue of the geographical names that are being mapped, and their quality, accuracy, and scale of mapping. Each of these concerns is addressed below.

The first issue (of identification of areas to map in preparation for disaster) has been covered in depth elsewhere with respect to 'traditional' GIS (Tomaszewski, 2015), but many neogeographers (represented by HOT) do not have experience as cartographers and GIS practitioners, nor as anthropologists (which creates issues for place-name mapping, as described below). As such, the insights contained in a text like *GIS for Disaster Management* may be overlooked by non(GIS)-experts. There are three possible ways to identify and target areas for future disaster mapping: 1) by targeting those areas already affected (blackboxing), 2) by autocorrelation (where something has already happened it is more likely to happen again in that location), and 3) by blanket coverage (to the extent that it is possible to map everything in a given area). Targeting areas already affected, or in the process of being affected, by a natural disaster makes sense because it is a proactive response; the same argument can be made for autocorrelation. However, both can lead to places being blackboxed. What this means is that mapping, led by outsider-trained locals, happens because the place is disaster-ridden, leading to an assuaging of guilt should the mapping take place. It can always be justified, so the argument goes, because it is free and it cannot possibly do harm to have such a base map in place. But mapping undertaken in response to a (future) disaster will be very different from that undertaken for play, for the placement of social services, or for the development of local and indigenous tourism, to name just three examples. The latter (tourism) can be negatively affected if an algorithm automatically associates a place with disaster. Increased effort after the fact only strengthens the association (and the associated, but now morally suspect, map).

Blanket coverage was given as a rationale by one proactive mapper with whom I spoke; they cited the entirety of 'Southeast Asia' as the target area! This is (obviously) a vast area and one can only surmise that the ambition of certain (western) activists outstrips their potential. What would the mapping of all of Southeast Asia on OpenStreetMap entail? It would require not only massive but also highly coordinated effort to carry off such a project, and it would necessitate taking seriously issues of scale, intensity, and coverage of the mapping effort. This would, in turn, require expertise from outside organizations—a category of which, it is worth pointing out, HOT is a member as well, having 'outsider' status to most local (usually student) communities. There is a certain hubris involved in thinking that goodwill and spotty local support could drive any kind of effective, rigorous, large-scale mapping effort such as that alluded to by the activist noted above. It is this kind of overweening pride that was the impetus for colonial mapping efforts in the New World, for example. Replace 'new world' (America) with a 'new' new world (Southeast Asia, in need of help according to western activists) and one has a newly filled in OpenStreetMap filled with locally defined but outsider-driven fantasy spaces of a western mind that sees the developing world as forever embattled, 'fallen', and on the verge of disaster.

This relates to the important issue of place-name mapping, which requires rigor and, as with disaster mapping, a carefully thought-out plan that needs to take into account scale, intensity, and coverage. Place-name mapping needs local language experts, geographical knowledge, and the sort of long-standing relationships with indigenous populations that are only fostered by well-established anthropological practice (Thornton, 1997; Clifford and Marcus, 1986). After establishing sound contact, language proficiency, and cultural understanding, an outsider researcher (now with a certain amount of 'insider' credibility and status through participation in everyday life) might be ready to consider mapping local place names. This requires the following for each toponym: the name, its meaning, and its precise location and shape. To give some indication of the scale of effort and planning involved, McGill professor Ludger Muller-Wille mapped almost 8,000 place names in Nunavik, (northern) Quebec with teams of elders in fourteen communities working in a coordinated fashion over decades. Each mapped name needs to then be validated and verified before being entered into a relational and spatial database (GIS). What is different from the rigorous scenario described above is that the database in OpenStreetMap is not properly spatial (i.e. it is 'flat') and also that its efforts are driven by non-experts, which in this case is a problem. In many areas where effort is targeted by, for example, HOT, the geographical names are being mapped for the first time. If established place-name mapping procedures are not followed, the effort will ingrain inaccurate and patchy toponym coverage for ever—or at least until the underlying database is cleaned up. Far from being a useful interim effort, the VGI neogeographers of HOT are setting a dangerous precedent that actually lowers the bar rather than raises it.

The same activist who made the comment above about mapping all of Southeast Asia also commented (on Twitter) that 'crappy' mapping would be better than no mapping at all. This is most certainly not the case, for all of the reasons given above and also because it ignores basic best practice in geography, anthropology, and GIS for disaster mapping (Tomaszewski, 2015). It risks slotting marginal areas into so-called 'crappy' maps that will most likely never be used because the disaster that is allegedly waiting to happen is so far in the future, or of such a nature, that the map will for all intents and purposes be both out of date and useless. This is the final issue that any such mapping effort must grapple with: how to provide resources for updates to maps after the first effort. HOT is only concerned with first-effort mapping (which is patchy in any case) and does not consider how often, or how, updates to maps will be made. The assumption here parallels that of previous rounds of colonial mapping (e.g. in North America) that saw maps as 1) objective, 2) all-encompassing, and 3) timeless. In other words, maps tend to 'freeze' knowledge in the time and place at which it was mapped. Furthermore, they tend to serve the interests of their makers (i.e. the funders, or bodies providing the economic impetus). As we have explored throughout this book, geographical names, as represented on maps, evolve through time and represent things at only one level (scale). New geographical names are constantly coming into existence or dropping out of existence, and these facts need to be taken into account. One cannot assume that, through some rapid-fire consultation and training process, mappers can simply parachute their way through a (re)mapping of the entire world. To do so would be to exhibit the same kind of hubris as colonial mappers of all stripes: imperial, anarchist, activist, progressive ... . The well-meaning (colonial) mapper of old is alive and well, and wearing the banner of progressive change—as has been the case from at least the time of Columbus.

Place branding

I shift gear now to explore a topic of great fascination for geographical names, that of place branding through such channels as social media and Twitter hashtags. Tourism companies have always had to grapple with brands, whether of the companies themselves or of the locations and areas they serve as target markets and destinations (Hanna and del Casino, 2003). Social media sites can help tourism targeting through the use of tags, which are neither simple labels nor unsophisticated memes. A hashtag (or tag) is often preceded by the '#' symbol, which is recognized by sites like Twitter and Facebook as preceding the tag and distinguishing it as such for both site and user. A tag is actually a small piece of metadata for the creation of what have been called 'folk taxonomies' or folksonomies, which are taxonomies created through the bottom-up participatory processes that Twitter hashtags and other social media activity represent (Smith, 2008).

Quantitatively, tags operate on a long-tail model, meaning that the frequency of the most popular tags attached to a resource (tweet, link, video, blog post, or anything with a Uniform Resource Locator, or URL) far outstrip the frequency of the less popular tags (Smith, 2008, page 53). The result is a 'name,' or new category in a folksonomy, for that resource. For our purposes, proper names and categories can be treated as equivalent, as confirmed by a range of philosophers who disagree on much else (Jubien, 2009; Fitch, 2004; Evans, 1982; Kripke, 1981). Thus, the creation of folksonomies through tags generally results in the highest frequency tag becoming the name for the 'thing', which is a resource located on the World Wide Web. For our purposes in this section we examine tags that have geographical content or are attached to resources with geospatial aspects. Admittedly, up to 80 percent of the content of the web has a spatial component, making the 'geo' aspect of the web (or geoweb) its 'killer app' (Scharl and Tochtermann, 2007). This makes it all the more important to have a good handle on what is occurring when objects are named through tagging and folksomony creation.

The branding of place increasingly forms part of place identity, especially urban places (Kavaratzis and Ashworth, 2005). A very interesting new way of branding place has arisen through VisitBritain's use of the #greatnames tag for use by Chinese tourists to London and the UK to rename their surroundings according to tourist sensibilities and language constructs not captured by existing local place names. The result has been descriptive-sounding names translated from the Chinese, as described in a video celebrating #greatnames:

> For centuries, the British roamed the world, slapping English names on just about everything. So when it came time to promote Great Britain in China, we thought the Chinese might like to return the favour, with great Chinese Names for Great Britain.
>
> (Visit Britain, 2015)

Thus, the well-known longest named place in Wales is called "Word-Puzzle town"; The Mall (in London) is called "Queen's Driveway"; Sherwood Forest is called "Forest of Chivalrous Thieves"; The Shard is called "Tower to Pluck the Stars"; Glen Coe is called "Grim but Colourful Valley"; Saville Row is called "Street for the Tall, Rich, and Handsome" (Visit Britain, 2015). Here, we have the replacement of proper names with Fregean names, in that translated Chinese noun phrases have replaced the more condensed English proper names. The latter have the benefit of years of transmission across generations, whose interaction and communication with the places and their proper names has resulted (for the English speaker) in an accumulation of associated encyclopedic knowledge. The Chinese, without the benefit of such accumulated knowledge resulting from something like Kripke's 'chains of communication', must resort to more Fregean descriptive naming practices for the production of their #greatnames neogeographies on the 'geoweb.'

The video goes on to note that an online mapping platform can be used to add new names:

> For places not listed on the [visitbritain.com] site, namers were encouraged to personally visit the location .... to upload their submissions to Weibo, along with proof of visit. A Chinese Beatles fan named 'You Run' grabbed the opportunity to name Beatles landmarks in London, resulting in the renaming of Abbey Road to "Love and Goodbye Road."
>
> (Visit Britain, 2015)

Overt referencing (with a heavy dose of humor) of colonial naming practices now being reversed by an implied Chinese colonization of the world is an innovative reframing of the idea of place branding. It upends an underlying anxiety about Chinese visitors not 'getting' English names for London places and a consequent (implied) (re)colonization of the city. This clever strategy both reduces the anxiety through humor and the possibility of increased tourism revenues and creates a potential goldmine of new names that open a window on to how outsiders see London.

With #greatnames, a tag has been 'pre-given' by Visit Britain for others to use and, as such, it is not a great example of a 'proper' folksonomy. The latter are generally driven from the bottom up, spontaneously and without external input. An academic paper on self-plagiarism, uploaded to a site like Academia.edu, for example, would most likely gain tags like #selfplagiarism or #Bauman because this is what the paper is about (see Lusher, 2015). In contrast, #greatnames is an attempt to rebrand the city of London and at the same time promote the city and increase the frequency with which Chinese tourists visit it.

#TelAvivSurSeine offers another fascinating example of neogeographical tagging for a place-branding effort, though in contrast to #greatnames, #TelAvivSurSeine would seem to have somewhat sinister connotations, especially for those opposed to the very idea of having a fake Tel Aviv on the banks of France's river Seine (Jenne, 2015, page 24). The one-day festival is part of a larger set of events collectively referred to as 'Paris Plages.' Protesters against the Tel Aviv part of this larger event were met by their alter egos across the river sporting the counter-moniker 'Gaza Plage,' accusing the other (Tel Aviv) side of supporting terrorism. It is not such a stretch to appreciate why some would take offense, but the authenticity of a Tel Aviv Sur Seine is akin to that of the Eiffel Tower replica found in Las Vegas. It is such a pale imitation that it cannot really be seen as a threat to the real thing. This is precisely what is meant by the word 'post-foundational' found in the title of the present work: the foundation upon which the name rests is in some sense illusory or less than stable, because the real-world referent of the name is constantly shifting and evolving new representations and aspects.

However, the protesters do have a point: the words Tel Aviv surely represent something real—and terrifying—to the residents of Gaza and their supporters. This is true despite Tel Aviv's reputation as a progressive place with a history of real opposition to Israel's activities, including the 2014 war in Gaza. But the mayor of Paris feels justified in maintaining Tel Aviv Sur Seine because of recent terrorist events in Paris:

> After the attacks on *Charlie Hebdo* and a kosher supermarket in Paris that left 17 dead in January, Manuel Valls, the Prime Minister, launched a three year action-plan against racial hatred. This week Mr Valls gave his 'full support' to Mayor Hidalgo's project.
>
> (Jenne, 2015, page 25)

Needless to say, the geopolitics of neogeographical, post-foundational place branding is fascinating and complex, with colonial overtones and implications. For Chinese tourists to London, #greatnames would seem to be less contentious than #TelAvivSurSeine. But the anxieties underlying #great-names are just as great if we consider the upsurge in China's economy and the resulting purchasing power of the average Chinese tourist. At the time of writing, the name China represents not only great hope for the world economy but also fears that China will be 'in control' or somehow not act responsibly, as actions around the recent 'black Friday' stock market crash might indicate. We both need the Chinese state and its travelers to hold things in place at a time of economic crashes, austerity, and uncertainty and fear their power. Ironically, the Chinese, known for their capacity to save, are now harbingers of a new consumerism which, brought to Britain, is threatening to rename (albeit humorously) the very place the tourism industry seeks to benefit, namely, Britain.

Mapping the flaneur

Names, in the flaneuristic senses in which Gregory (1994) and Benjamin (1999) evoke them, are well worth further exploration. In the post-foundational world of naming that exists on the geoweb, with tags, links, and mapping platforms coming to define the texture and geographies of the world, the rebranding of names can provide some backing for the claim put forward in this book that, in a Kripkean sense of possible worlds, 'another world is possible.' This view of names in reconfiguring geographies of everyday lives and life-worlds is less about the 'proper' aspect of names and more about 'noun phrases.' As Jubien (2009) and Eades (2015b) point out, matter in space-time can be discretized in infinite ways to produce new objects, relations, and memes for (re)naming the world. Tools such as GIS, the geoweb, GPS, and human cognition and cognitive maps aid in the construction of radically new NTNs for creating 'other' (possible, alternative) worlds.

By becoming a tourist in one's own backyard, psycho-geographical mappings of one's (urban) surroundings using cameras, GPS, and new mapping platforms (e.g. Google Earth) (Kingsbury and Jones, 2009; Eades, 2010), the idea of #greatnames takes on new potential significance for individuals. Those empowered with the equipment and (naming) tools can explore, poetically, new noun phrases for describing the city or immediate surroundings. This way of seeing the world, psycho-geographically, is part of a tradition of Situationist-influenced flaneurie that has a decades-long lineage, having begun with Guy Debord (who was in turn influenced by Walter Benjamin) who passed the torch on to cultural descendants like Self and Steadman (2007), O'Rourke (2013), and Wark (2011). Though Wark (2011, pages 165–166) differentiates Debord's dérive and the psycho-geographer from the flaneur in the following way, we maintain that no such clear-cut binary, for the purposes of neogeographical naming, is moot: "The dérive is different from the amblings of the flaneur, a more exclusively masculine figure for whom the street is to be seen as a thing apart, rather than a succession of atmospheres and adventures to participate in."

Bourgeois man/flaneur and situationist psycho-geographer alike are invited to explore some of the findings of the present work. Mapping, naming, and alternative boundary construction can be explored by re-envisioning language and mapping to recreate surroundings using methods and philosophies outlined here. A spirit of play abounds in both (flanerie and psycho-geography), and in the spirit of the #greatnames, new mappings are given 'permission' to playfully and humorously subvert established boundaries, languages, and names.

In 2008 I undertook a program of systematic wandering in Montreal, not far from my home, in order to complete a somewhat manic spatial renaming of my surroundings. It involved the translation of Charles Baudelaire's poem "Les Hiboux," each line of which was 'mapped' onto a painting of an owl by Jean-Paul Riopelle. The latter was used as an overlay on a map of Montreal, and various 'matches' were found between Riopelle's brushstrokes and the streets and roundabouts of the suburbs and surroundings of the city centre. I endeavored to walk those brushstrokes interpretively, documenting anything resembling or evocative of Baudelaire's owls along the way (Eades, 2009).

Separate lines of the translated Baudelaire poem acted like Fregean noun phrases for (re)translating the city into artistic terms. New to the city, I was in the act of translating myself to new surroundings, and what better way to do so, as an English speaker in a bilingual (French/English) city, than to shake up some of my monolingual assumptions? The noun-phrase lines were translated directly from Baudelaire's poem

> Les Hiboux
> Sous les ifs noirs qui les abritent,
> Les hiboux se tiennent rangés,
> Ainsi que des dieux étrangers,
> Dardant leur oeil rouge. Ils méditent.
>
> Sans remuer ils se tiendront
> Jusqu'à l'heure mélancolique
> Où, poussant le soleil oblique,
> Les ténèbres s'établiront.
>
> Leur attitude au sage enseigne
> Qu'il faut en ce monde qu'il craigne
> Le tumulte et le movement;
>
> L'homme ivre d'une ombre qui passé
> Porte toujours le châtiment
> D'avoir voulu changer de place.
>   (Baudelaire, 1861, p. 104)

I have roughly translated Baudelaire's poem as follows: Darkness under yews guards/ The owls' self-restraint/ Thus from strange gods/ Eyes darting red. They contemplate.// Without stirring they hold themselves/ Until the sad hour/ That, pushing the slanting sun,/ The darkness settles in.// Their wise bearing shows/ What it takes in this world that he fears/ The uproar and the burst;// The drunken shadow of a man/ Opens the doors of punishment/ For having wanted to move.

Note that the poem is not explicitly spatial, adding further to the challenge of translating it into the materiality of a dérive. But the poem's emotional content more than compensates something compatible with a Debordian sensibility. The dérive technique is defined by Debord (1981, p. 50) as an activity in which

> one or more persons during a certain period drop their usual motives for movement and action, their relations, their work and leisure activities, and let themselves be drawn by the attractions of the terrain and the encounters they find there. The element of chance is less determinant than one might think: from the dérive point of view cities have a psychogeographical relief, with constant currents, fixed points and vortexes which strongly discourage entry into or exit from certain zones.

In this project I did indeed drop my usual engagements, social and work activities, in order to psycho-geographically map Montreal. This involved a quite convoluted methodology of, as mentioned above, aligning a Jean-Paul Riopelle painting with the street grid of Montreal which served as a rough heuristic for guiding my movements over the course of fourteen days of psycho-geographical walking (one for each line in the poem). My self-assigned 'task' for each day was to 'capture' a representation of that day's line of poetry using a disposable camera. I had, at the time, read Wittgenstein's *Philosophical Investigations*, but the work was written up using more of a Deleuzean/Debordian framework. In retrospect, the focus on visual representations of text-lines from a poem is very Wittgensteinean in orientation, and served as an early portent of the material presented in this book.

The final 'move' in my project to 'map the flaneur' (myself as Baudelaire/ Riopelle) in Montreal was to geotag (using GPS) and post the fourteen photos (one for each of the poem's lines) to the social networking site Flickr (see Figure 6.1), which allows users to make maps of the locations of such geotagged photos. This 'final' map of my dérive excursions over two weeks represented a new topological space for viewing the poem (Eades, 2009). This map was viewable as an array of points across Montreal with rough correspondence to some of the lines of Jean-Paul Riopelle's owl (see Figure 6.2).

The inclusion of Flickr in the project's methodology is a neogeographical maneuver, designed to allow for the display of a topology of being for the poem constrained not only by the imagination but also by the geospatial technological underpinnings of the platform/site (Flickr). These are what are called, in psycho-geographical parlance, 'creative constraints' for generating new insights into space and spatiality, especially in cities. This use of

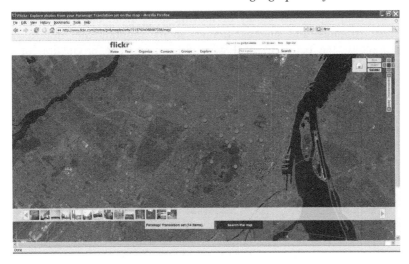

Figure 6.1 Geotagged Flickr photos
Source: Map data ©2015 Google.

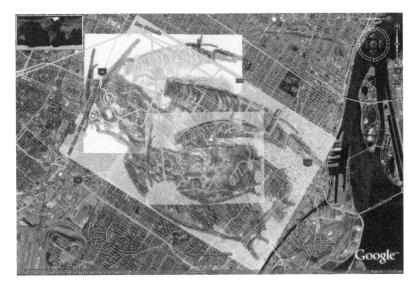

Figure 6.2 Georeferenced Riopelle owl

geospatial technologies is part of what I described in *Maps and Memes* as a method of healing in the wake of colonial cartographic legacies and, for indigenous peoples, more damaging legacies of displacement and incarceration in the very lands and life-worlds from which their ancestors derived livelihood and meaning (Eades, 2015b). The use of creative constraint combined with geospatial technology is generative of new insight and connection and, as such, represents a way of moving away from damaging historical erasures, ruptures, and silences. It does so by adopting the rupture as its methodology, the better to deal with it head-on. The production of new names for spaces is given a visual and thus more concrete edge and the immediate, tangible, and hands-on results give it potential therapeutic value.

As we move towards some conclusions, the final word of this chapter is that for neogeography and psycho-geography there is nothing to conclude. Flanerie is about moving on to the next attraction, emotion, or mood. For now we must let the flaneur go about his/her merry/melancholy way, to consider some fixed points in various landscapes we have named along the way.

References

Adler, Ron and Srebro, Haim. 2015. Boundary Surveying in the Middle East, in Monmonier, Mark (ed.). *The History of Cartography*. Vol. 6, *Cartography in the Twentieth Century*. Chicago: University of Chicago Press.

Akerman, James R. (ed.). 2008. *The Imperial Map: Cartography and the Mastery of Empire*. Chicago: University of Chicago Press.

Althusser, Louis. 1977. *Lenin and Philosophy and Other Essays*. London and New York: New Left Books.

Austin, Ian. 2012. Coming Soon, Google Street View of a Canadian Village You'll Never Drive To. *New York Times*. August 22.

Barr, James. 2011. *A Line in the Sand: Britain, France and the Struggle that Shaped the Middle East*. London: Simon & Schuster.

Baudelaire, C. 1861 [Éditions Gallimard, 1996]. *Les Fleurs du Mal*. Paris: Gallimard.

Benjamin, Walter. 1999. *Illuminations*. London: Pimlico.

Bergson, Henri. 1990. *Matter and Memory*. New York: Zone.

Borges, Jorge Luis. 2000. *Selected Poems*. New York: Penguin.

Bowcott, Owen and Ball, James. 2014. Revealed: how UK exploits snooping laws. *The Guardian*. 14 June.

Casey, Edward. 2002. *Representing Place: Landscape Painting and Maps*. Minneapolis: University of Minnesota Press.

Casey, Edward. 1997. *The Fate of Place: A Philosophical History*. Berkeley and Los Angeles: University of California Press.

Casey, Edward. 1993. *Getting Back Into Place: Toward a Renewed Understanding of the Place-World*. Bloomington and Indianapolis: Indiana University Press.

Clifford, James and Marcus, George E. (eds). 1986. *Writing Culture: The Poetics and Politics of Ethnography*. Berkeley and Los Angeles: University of California Press.

Cockburn, Patrick. 2015. *The Rise of Islamic State: ISIS and the New Sunni Revolution*. London and New York: Verso.

Cockburn, Patrick. 2014. "Destroyed in a Second": Entire Neighbourhoods Demolished by Syrian Army. *The Independent*. January 31.

Collignon, Beatrice. 2006. *Knowing Places: Innuinnait, Landscapes and Environment*. Calgary: CCI Press.

Cope, Meghan and Elwood, Sarah (eds). 2009. *Qualitative GIS: A Mixed Methods Approach*. London: Sage.

Debord, G. 1981. Theory of the Dérive, in Knabb, Ken (ed.). *Situationist International Anthology*. Berkeley: Bureau of Public Secrets. 50–54.

Deleuze, Gilles. 1990. *Bergsonism*. New York: Zone.

Eades, Gwilym. 2015a. The Hubris of Proactive Disaster Mapping. *Place Memes* (blog). http://place-memes.blogspot.co.uk/2015/05/the-hubris-of-proactive-disaster-mapping.html (accessed August 25, 2015).

Eades, Gwilym. 2015b. *Maps and Memes: Redrawing Culture, Place, and Identity in Indigenous Communities*. Montreal and Kingston: McGill-Queen's University Press.

Eades, Gwilym. 2015c. Hardy vs the Windfarms: Tess, Necessity, and the Geopolitics of the Anthropocene. Presentation to the Royal Geographical Society, Exeter, UK.

Eades, Gwilym. 2010. *An Apollonian Appreciation of Google Earth*. Geoforum. 41(5). 671–673.

Eades, Gwilym. 2009. Mapping the Flaneur: A Geospatial Translation of Charles Baudelaire's "Les Hiboux". Presentation to the Canadian Association of Geographers meeting, Carleton University, Ottawa, Ontario, Canada.

Eades, Gwilym and Zheng, Yingqin. 2014. Counter-Mapping as Assemblage: Reconfiguring Indigeneity, in Doolin, Bill, Lamprou, Eleni, Mitev, Nathalie and McLeod, Laure (eds). *Information Systems and Global Assemblages: (Re)configuring Actors, Artifacts, Organisations*. Heidelberg: Springer.

Economist, The. 2014. Missing Maps and Ebola (Daily Chart). November 13. www.economist.com/blogs/graphicdetail/2014/11/daily-chart-8 (accessed August 27, 2015).

Edney, Matthew H. *Mapping An Empire: The Geographical Construction of British India, 1765–1843*. Chicago: University of Chicago Press.

Evans, Gareth. 1982. *The Varieties of Reference*. Oxford: Oxford University Press.

Fisk, Robert. 2015. This Latest Isis Snuff Movie is the Stuff that Nightmares—and Dreams—Are Made Of. *The Independent*. August 17.

Fitch, G. W. 2004. *Saul Kripke*. Chesham: Acumen.

Gelling, Margaret and Cole, Ann. 2000. *The Landscape of Place-Names*. Stamford: Shaun Tyas.

Gibson, William. 2007. *Spook Country*. London: Penguin.

Goodchild, Michael F. 2007. Citizens as Sensors: The World of Volunteered Geography. *GeoJournal*. 69(4). 211–221.

Gregory, Derek. 1994. *Geographical Imaginations*. Cambridge, MA and Oxford: Basil Blackwell.

Guardian, The. 2014. Mark Carney: Most Fossil Fuel Reserves Can't Be Burned. October 13.

Hanna, Stephen P. and Del Casino, VincentJ. (eds). 2003. *Mapping Tourism*. Minneapolis: University of Minnesota Press.

Hardy, Thomas. 1978. *Tess of the D'Ubervilles*. London: Penguin.

Harper, Tom and Owen, Jonathan. 2014. Google privacy law 'means total rethink of basic freedoms'. *The Independent*. May 31.

Hill, Linda L. 2006. *Georeferencing: The Geographic Associations of Information*. Cambridge, MA: MIT Press.

Hu, Tung-Hui. 2015. *A Prehistory of the Cloud*. Cambridge, MA: MIT Press.

Huggett, Richard John. 1997. *Catastrophism: Killer Asteroids in the Making of the Natural World*. London and New York: Verso.

Husain, Aiyaz. 2014. *Mapping the End of Empire: American and British Strategic Visions in the Postwar World*. Cambridge, MA: Harvard University Press.

Jenne, Amelia. 2015. Tel Aviv Sur Seine: Arab-Israeli Debate Muddies the Party. *The Independent*. August 14.

Johnson, Leslie Main. 2010. *Trail of Story, Traveller's Path: Reflections on Ethnoecology and Landscape*. Edmonton: Athabasca University Press.

Johnson, Leslie Main and Hunn, Eugene S. (eds). 2010. *Landscape Ethnoecology: Concepts of Biotic and Physical Space*. New York and Oxford: Berghahn.

Jones, Steven E. 2013. *The Emergence of the Digital Humanities*. London: Routledge.

Jubien, Micheal. 2009. *Possibility*. Oxford: Oxford University Press.

Kavaratzis, Mihalis and Ashworth, G. J. 2005. City Branding: An Effective Assertion of Identity or a Transitory Marketing Trick? *Tijdschrift voor Economische en Sociale Geografie*. 96(5). 506–514.

Kingsbury, Paul and Jones, John Paul III. 2009. Walter Benjamin's Dionysian Adventures on Google Earth. *Geoforum*. 40(4). 502–513.

Kripke, Saul. 1981. *Naming and Necessity*. Malden: Blackwell.

Lawrence, Vanessa. 2013. Personal communication.

Leszczynski, Agnieszka. 2014. On the Neo in Neogeography. *Annals of the Association of American Geographers*. 104(1). 60–79.

Lewis, Simon L. and Maslin, Mark A. 2015. Defining the Anthropocene. *Nature*. 519(7542). 171–180.

Lusher, Adam. 2015. Zygmunt Bauman: World's Leading Sociologist Accused of Copying his Own Work. *The Independent*. August 20. www.independent. co.uk/news/uk/home-news/zygmunt-bauman-worlds-leading-sociologist-accused-of-copying-his-own-work-10464486.html?origin=internalSearch (accessed August 28, 2015).

MailOnline. 2013. Thomas Hardy Crowds' Madding Fury over Plans to Build 400ft Wind Turbines on Countryside which Inspired Famous Novelist. www.dailymail.co. uk/news/article-2487008/Thomas-Hardy-crowds-madding-fury-plans-build-400ft-wind-turbines-countryside-inspired-famous-novelist.html#ixzz3jAYV0kjN. November 4 (accessed August 18, 2015).

Malpas, Jeff. 2006. *Heidegger's Topology: Being, Place, World*. Cambridge, MA: MIT Press.

Nagel, Thomas. 2002. *Concealment and Exposure & Other Essays*. Oxford: Oxford University Press.

Michael, Chris. 2014. Missing Maps: Nothing Less Than a Human Genome Project for Cities. *The Guardian*. October 6.

Natural Resources Canada. 2015. *The Atlas of Canada*. www.nrcan.gc.ca/earth-scien ces/geography/atlas-canada (accessed August 24, 2015).

O'Rourke, Karen. 2013. *Walking and Mapping: Artists as Cartographers*. Cambridge, MA: MIT Press.

Pasquale, Frank. 2015. *The Black Box Society: The Secret Algorithms That Control Money and Information*. Cambridge, MA: Harvard University Press.

Peterson, Michael. 2014. *Mapping in the Cloud*. New York: Guilford.

Pizzi, Michael. 2015. Egypt to Imprison Journalists who Report False Death Tolls. *Al Jazeera America*. http://america.aljazeera.com/articles/2015/7/6/egypt-to-jail-journalists-who-report-false-death-tolls.html (accessed April 10, 2016).

Pomerantz, Jeffrey. 2015. *Metadata*. Cambridge, MA: MIT Press.

Ramm, Frederik, Topf, Jochen and Chilton, Steve. 2010. *OpenStreetMap: Using and Enhancing the Free Map of the World*. Cambridge: UIT Cambridge.

Scharl, Arno and Tochtermann, Klaus (eds). 2007. *The Geospatial Web: How Geo-browsers, Social Software and the Web 2.0 are Shaping the Network Society.* London: Springer-Verlag.

Self, Will and Steadman, Ralph. 2007. *Psychogeography.* London: Bloomsbury.

Sengupta, Kim. 2015. Turkey's Buffer Zone in Syria: Self-defence—Or Just Anti-Kurd? *The Independent.* August 21.

Shifman, Limor. 2014. *Memes in Digital Culture.* Cambridge, MA: MIT Press.

Sieber, Renee. 2006. Public Participation Geographic Information Systems: A Literature Review and Framework. *Annals of the Association of American Geographers.* 96(3). 491–507.

Smith, Gene. 2008. *Tagging: People-Powered Metadata for the Social Web.* Berkeley: New Riders.

Stern, Pamela and Stevenson, Lisa (eds). 2006. *Critical Inuit Studies: An Anthology of Contemporary Arctic Ethnography.* Lincoln and London: University of Nebraska Press.

Thornton, Thomas F. 1997. Anthropological Studies of Native American Place Naming. *American Indian Quarterly.* 21(2). 209–228.

Thornton, Thomas F. (ed). 2012. *Haa Leelk'w Has Aani Saax'u/Our Grandparents Names on the Land.* Seattle, London and Juneau: University of Washington Press/ Sealaska Institute.

Tomaszewski, Brian. 2015. *Geographic Information Systems for Disaster Management.* Boca Raton: CRC Press.

Turner, Andrew. 2006. *Introduction to Neogeography.* San Sebastopol: O'Reilly.

Vincent, James. 2014. Snooping on Social Media is Legal, Says Anti-terror Chief. *The Independent.* June 18.

Visit Britain. 2015. Visit Britain: Great Chinese Names for Great Britain. https://vim eo.com/128456211 (accessed August 28, 2015).

Wark, McKenzie. 2011. *The Beach Beneath the Street: The Everyday Life and Glorious Times of the Situationist International.* London: Verso.

Wellen, Christopher. 2008. Ontologies of Cree Hydrography: Formalization and Realization. Unpublished MSc Thesis. Montreal: McGill University, Department of Geography.

Zizek, Slavoj. 1993. *Tarrying With the Negative: Kant, Hegel, and the Critique of Ideology.* Durham, NC: Duke University Press.

7 Toward a geographical name-tracking network

This book has used a hybrid descriptive-analytical definition of names (combining noun phrases and proper names) with a name-tracking network framework as a toolset for examining the nature of the geographical name over time and space. Names have been demonstrated to have evolved in significant ways from more descriptive indigenous and religious naming practices that retain influences prior to contact with colonizing (outside) forces. With contact and hybridization of culture that evolution of the name has moved toward a more political, counter-mapped, and post-foundational geography for resisting, adopting, and adapting to hegemonic cultural influences.

This concluding chapter serves to explore some of the philosophical and poetic conclusions that can be made from discovering that the very nature of the geographical name has evolved over time and space.

The power of names

What or who names? It makes a difference. Monmonier (2006) has claimed that maps "name, claim, and inflame," but do maps really have this power? Assigning agency to names requires those agents to do such things as baptizing the previously unnamed by a new referring description or proper name. But, as with maps, we tend to forget that behind the name lies an interest in seeing that name survive. One could say this is the 'real' agency behind any name, the person or community that performs the act of naming (i.e. the baptism or original descriptive moment). What the name represents, and for whom, matters. Changing place-naming practices have often been driven by capitalistic and colonial imperatives which made the sovereign's representative in the colony (the explorer or the surveyor) responsible for naming. The representative acted as middleman, marketing the New World for an audience in the old world through the medium of the map (Yeazell, 2015). Both the middlemen, and their maps, can be said to be agents of the (geographical) name (Hornsby, 2011).

Figure 7.1 makes clear that changing practices are themselves middlemen in the 'grand scheme' of geographical names. Practices change, as do both names and their referents, creating a sort of moving baseline for the scholar of

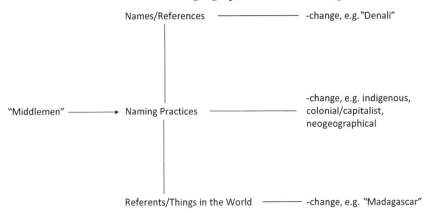

Names/References ────────── -change, e.g. "Denali"

"Middlemen" ───────▶ Naming Practices ────────── -change, e.g. indigenous, colonial/capitalist, neogeographical

Referents/Things in the World ───────── -change, e.g. "Madagascar"

Figure 7.1 Geographical naming practices

place names (Gelling, 1978). Berger (2002, page 19; Evans, 1982) looks, in an example of referent change, at the

> 'Madagascar case' due to Gareth Evans. According to him, 'Madagascar' was a native name for a part of the continent of Africa. Marco Polo erroneously took the natives to be using that name to refer to an island off the coast of Africa. Today the term is so widely used as a name for the island that this usage has overridden the earlier historical connection to the referent assigned to 'Madagascar' at the initial baptism.

We can now begin to sketch out a proper typology of geographical naming practices, using Figure 7.1 as a guide. This typology is derived from the 'middlemen' section of that diagram that refers to geographical naming practices and their changing nature from indigenous to colonial/capitalistic to neogeographical. This way of describing changing practices is not intended to map any idea of 'progress' to geographical names (Edney, 1993). Instead, as I spell out the typography of geographical names below (Table 7.1), keep in mind that traces of prior practice remain as both palimpsest and 'layer' for hybridizing aspects of geographical names and naming practices. Thus, indigenous practices can both inform and be replaced by colonial/capitalist systems; while neogeographical practices are in some ways a return to descriptive indigenous systems. In reality there is no neat way of separating these three kinds of discourse, but in scholarly practice we must attempt to make sense of the world through categorization and critical assessment (and questioning) of the archive at hand. This consists, in the present work, of a diverse array of documents, inscriptions, observations, databases, geographical names, and counter-mappings that strongly suggest such hybridities and layerings are very real indeed.

Indigenous names continue to be 'baptized' all over the world as demand in indigenous communities to both preserve traditional knowledge systems and

Table 7.1 Typology of geographical naming practices

Colonial place-naming practices (primarily satisficing, or S-type)
—ascriptive (not primarily descriptive) —eponymous (name-borrowing) —commemorative/historical (remembering 'home' places) —inscribed (made permanent on maps) —political/patronising (hierarchical)
Indigenous place-naming practices (primarily focal, or F-type)
—descriptive (ostensible focus on a feature) —action-oriented (linked to travel/events) —topological (linked across the landscape) —utilitarian (useful for wayfinding and location of resources) —performed (names uttered in interaction with the land)
Neogeographical place-naming (mock-focal, or mF-type)
—memetic/viral (horizontal spread of information) —descriptive (ostensible focus on feature/event) —ephemeral (short-lived) —social (multi-media) —networked (tagged)

make sense of rapid global (and climate) change persists. The appropriate typology here is one of what Berger calls focal-type (or F-type) naming, which is performative and, further, requires direct sensory focus on the thing being named (Berger, 2002, page 4). There is a distinct sense of acquaintance through direct perception and interaction with the (geographical) thing-in-the-world being named in indigenous practice. "Mount Denali, an Athabascan name used by generations of Alaska Natives that means 'the great one,'" according to Mufson (2015), and the mountain was named (ab)originally for its size and prominence in the landscape, and for its greatness as it exceeded surrounding mountains in magnitude. But the prior name was usurped for many years by the name Mount McKinley, an example of a colonial place-naming practice that Berger (2002) calls S- (for satisficing) type. S-type names satisfy a condition specified by some criteria, whether explicit or implicit. For example, it was implicit to the surveyor's craft in (re)naming features in North America, to pay lip-service to political power (i.e. that of president McKinley) at the imperial (American) centre to which the surveyor was often beholden (i.e. paid by). While much can be said for individual preferences in geographical naming styles, colonial geographical naming practices resonated with each other in certain respects, despite individual idiosyncrasy. Hornsby (2011, page 128) points out that James Cook named features during travel along coastlines, mixing proper names of prominent personages with those describing events or qualities of the landscape. At an extreme end was Holland, who named St John's and Cape Breton islands according to a strictly hierarchical schema of civic boundary and size corresponding to prominence of royal and political rank back in England. The largest civic units belonged (nominally) to king, queen and prince; while largest bays were named after most prominent (and individually favourable) politicians in the life of the surveyor (Hornsby, 2011). With the above in mind, therefore, Table 7.1 presents the typology.

Berger (2002) was discussing names in general, not specifically place names, but his logic applies especially well to the geographical case. The distinction between F-type and S-type geographical naming practices has fairly clear compatibilities with indigenous and colonial/capitalist naming practices, but there are obvious areas of overlap. This is especially true when discussing the transmission of names through time and across space, referred to earlier in this book as memetic intergenerational qualities of place names that are vertical (time) and horizontal (space). Berger (2002, page 14) refers to anaphoric-background (A-B) conditions for the transmission of names. A-B conditions indicate whether the origin of a name is an object, a fictional subject or person, a satisficing condition, or a mock-focal operation of some kind. The latter (mock-focal) method of transmission applies to all name transmissions after the initial baptism, when the object is no longer present or remembered directly by any living member of the community. 'Aristotle' could be one such case, as could arguably (in the geographical case) 'Atlantis' or 'Easthampton' (in England). These places do not exist in the present day, nor did they exist within living memory, but the names persist in discourse. In these cases, Berger (2002, page 15) argues, community belief in a satisfactory (as opposed to satisficing) description comes to bear on its validity.

So Aristotle can be described as the author of the *Nicomachean Ethics* so long as the community of practice believes this to be the case. One can similarly ascribe belief to the 'existence' of Atlantis (as a lost civilization), Easthampton (as a lost archaeological relic) or even the beauty (or non-beauty) of a place like London on the basis of community-led agreement on the subject. Due to overlooking the social and community-level aspects of geographical naming discourses it has taken the philosophical community some time and many wasted words to sort out some crucial points about belief. Kripke can be credited, however, briefly, with giving the social aspect of naming its due in *Naming and Necessity*, but many others failed to heed the call until Berger (2002) essentially sorted things out. Evans (1982), for example, despite mentioning community in the final chapter of his influential *The Varieties of Reference*, takes what Hanna and Harrison (2004) characterize as a hubristic approach—that is, one that simply reads meaning directly from the objective world without the intervention or bridging of human and community practices. The strength of their *Word and World* is the rigor with which they set out to exhaustively document this hubris and set out an alternative paradigm productive of what they call the name-tracking network, or NTN, which has formed a key theoretical part of the present work.

Indeed, I take up Hanna and Harrison's (2004) call with my own: to produce nothing less than a geographical NTN (or geo-NTN) that ties into the three main kinds of practices identified in the typology in Table 7.1. The last of these, the neogeographical, with its proliferation of tools, is posited as the most conducive to the production of such an archive that must, given current technologies, pay heed to current trends in interoperable GIS, metadata, big data, and online archiving systems. In the case of 'Denali'/Mount McKinley/ Denali, notational systems for tracking reference change (as seen in the top-right part of Figure 7.1) are of crucial importance. Metadata about place names whose reference (name) changes over time must reflect, somehow, the fact that the new Denali (without quotes to show that it is now primarily an inscribed rather than orally transmitted name) is a political move motivated by the president of the United States who is, ironically (but apparently without any intended irony), displacing a previous president's name from the tallest mountain in North America. It must also reflect the fact that the 'new' Denali is not the same as the old 'Denali'. The latter (in quotes) was part of an oral tradition that included names in stories handed down over generations in an area that has now come to be known as Alaska (a name itself derived from prior, oral, sources). Obama's was a politically calculated move, as he himself made no bones about revealing (Mufson, 2015).

The philosophy of names

Berger's (2002) notion of ascriptive transmission of proper names, in which the proper name (without linking description) is communicated from individual to individual across generations, is both astute and pertinent in the case of geographical names. For example, since a number of potential descriptions could be attached to the name 'Aristotle,' the transmission of the name itself is not dependent on any one description for it to be successfully preserved through time. This lack of descriptive content in name transmission is therefore ascriptive. It holds true for many post-indigenous (and post-foundational) geographical names as well, since, for example, a name like 'Paris' need not be attached to any one particular description (despite the urge to reduce the analysis of that name to something like the Eiffel Tower or similar object descriptions) for the proper name to be preserved through time across many generations. Colonial and Christian naming practices are intimately caught up in this kind of 'transmission' logic, whereby a condensation or crystallization of an entity (place or person) in the form of a proper name is more easily preserved than its associated descriptions (which after all are multiple and lose force over time). The name can stand in for the entity it 'labels' and serve as a place marker for identity (and essence) through the thick and thin of time.

Indigenous naming practices are quite different due to their performed and embodied nature. Indigenous geographical names, in particular, have been demonstrated to be descriptive, and those descriptions are handed down through the generations, sometimes in increasingly condensed forms or in conjunction with increasingly thick layers of colonial names alongside, or even hybridized with, the original descriptive content (Thornton, 2008; Basso, 1996).

The poetics of names

A crossover or hybrid style of naming is associated with poetics, as suggested by Wordsworth in his "Poems on the Naming of Places," in which unnamed places are given whimsical names based on emotion or other criteria (Wordsworth, 2008, pages 199–206). This is a hybrid of indigenous and neogeographical in the sense that names are suggested descriptively (indigenous) but not according to some immediate need, and almost whimsically (neogeographical). For example, the first of these poems is subtitled "It was an April morning: fresh and clear," proceeding to a series of observations about a place that ends up being named after a person (i.e. a proper name)

> To spur the steps of June; as if their shades
> Of various green were hindrances that stood
> Between them and their object: yet, meanwhile,
> There was such deep contentment in the air
> That every naked ash, and tardy tree
> Yet leafless, seemed as though the countenance
> With which it looked on this delightful day
> Were native to the summer.
> […]
> I gazed and gazed, and to myself I said,
> "Our thoughts at least are ours; and this wild nook,
> My EMMA, I dedicate to thee."
>
> (Wordsworth, 2008, page 200)

The observations and descriptions serve as ad hoc descriptions gathered by the poet to be attached to the place that is named eponymously, in this case Emma's Dell. This "wild nook" is a place where the poet can keep his thoughts to himself, and think as well upon one (Emma) with whom he would most like to share those thoughts. Only the shepherds are privy to such wild knowledge and, indeed, to this one-off name that appears only in the poet's thoughts. It is now, of course, also inscribed for all time in the canon of world literature, but that is no matter. For there is no real Emma's Dell. The point of the poem is how place and nature come to stand in for our passion, individuality, and deepest thoughts. The whole poem is a set of noun phrases that lead descriptively to the ad hoc eponymous name, proper (Wordsworth, 2008, page 200).

The second of Wordsworth's (2008, page 201) poems on naming places is also eponymous, having the sub-title "To Joanna." The name Joanna is actually chiseled into the rock, an inscription of place in the most literal sense, completed by a fanatically devoted lover of the woman in question, who chisels the name of his loved one into a rock overlooking a scene of profound and sublime beauty

And when we came in front of that tall rock
Which looks towards the East, I there stopped short,
And traced the lofty barrier with my eye
From base to summit; such delight I found
To note in shrub and tree, in stone and flower,
That intermixture of delicious hues,
Along so vast a surface, all at once,
In one impression, by connecting force
Of their own beauty, imaged in the heart.
[...]
In memory of affections old and true
I chiseled out in those rude characters
Joanna's name upon the living stone.
                    (Wordsworth, 2008, pages 202–203)

These places in Wordsworth's poems are not real in any literal sense. They are metaphors for fleeting beauty and love, but they represent the universality of those aspects of humanity through the medium of landscape and the outer, wild, world at large. As such, the noun phrases about nature produced by Wordsworth in these poems, each of which ends in a name (or an allusion to one), send a message to the reader about language itself, its connection to the world, and practices for representing it. Wordsworth chooses his language carefully, often solipsistically, and it is often about the Self.

In "There is an Eminence,—of these our hills" (the subtitle of Wordsworth's third poem about named places), no name is given other than a place holder for

She who dwells with me, whom I have loved
With such communion, that no place on earth
Can ever be a solitude to me,
Hath said, this lonesome Peak shall bear my Name.
                    (Wordsworth, 2008, page 203)

The use of capital letters here indicates that Wordsworth is speaking of a universal operation for naming places based upon their Eminence; that of the person doing the naming is, of course, implied, and stands in for the romantic lover of nature who views the (named) peaks, valleys, streams, trees, and flowers that stand for nature in all its metonymical heterogeneity and plenitude. Only in a metaphorical sense, then, can one posit the existence of a geo-NTN in a Wordsworthean sense, given as he is to relying upon noun phrases for describing nature. If noun phrases can be tracked (i.e. through poetry), then it is poetic practice itself, in all its varied forms, that can be said to 'produce' such an NTN.

Hardy's Wessex, as described in the novels and poems, offers another example of poetic naming that, despite the correspondences with the real world noted above, have created a fictional geographical NTN for tracking fictional names through time. The ideas associated with 'Wessex,' 'Tess,' 'Jude,' and other names will continue to resonate through time and across space in new, politically oriented forms, as well as in fictional influences (Hall, 2015) and cultural productions (e.g. on television). Hardy's poems about Wessex are less literally about naming—or, rather, less about proper names—than about Fregean description. For example, the poem "Neutral Tones" (Hardy, 2009, pages 1–2), from the *Wessex Poems*, is about place, personified in names (noun phrases) of bleak nature; and it is about a person, a lost lover perhaps, painted in the tones alluded to in its title

> The smile on your mouth was the deadest thing
> Alive enough to have the strength to die;
> And a grin of bitterness swept thereby
> Like an ominous bird a-wing … .
>
> Since then, keen lessons that love deceives,
> And wrings with wrong, have shaped to me
> Your face, and the God-cursed sun, and a tree,
> And a pond edged with greyish leaves.

Such a poem leaves a bitter taste in the reader's mouth, for it is about the deadness of a love and thus of a (metaphorical and real) place in the writer's heart. That metonymy and metaphor are deftly used by Hardy to shape his message is obvious; that they form part of a geo-NTN for producing poetic discourses of place is less so. Nineteenth-century writers were especially deft at such practices and their legacies are still felt today, not only in place names but also in styles of writing and relating to nature and the environment. Jane Austen's names (Doody, 2015), for example in *Mansfield Park* and *North-anger Abbey*, are similarly related to real and imagined places roamed by the fictional characters. *Emma* herself roams a set of very real places in Surrey in the midst of her fictional doings. In the twentieth century, Marechal's *Adam Buenosayres* (Marechal, 2014) is an example of writing in the tradition of Joyce's *Ulysses* for exploring how people are like places, organic wholes experienced from the inside in the course of a day or a similarly short period of time. The compression of action into a short time span allows the idea of place as character to be explored in more depth and, in Marechal, we have a proper (fictional) name (Adam) attached to a real place (Buenosayres) for the production of a metonymical and geographical NTN for tracking noun phrases that produce meaning around the capitol city of Argentina. Many other examples similar to this (but not always of such high quality) abound in world literature.

Kripke's (2011) "A Puzzle About Belief" is his most explicitly geographical work, and the introduction to the present work alluded to it indirectly. We have argued extensively through several chapters now that there is a strong link between belief and geographical names. Belief in a geographical sense (i.e. about places and spaces) is heavily influenced by names for things-in-the-world (referents), and by place-naming practices of the time in which specific geographical names 'baptize' entities (real or imagined) and discourses for contextualizing their present or contemporary utterance. Kripke (2013) and Jubien (2009) are both at pains to provide a philosophical framework for 'dealing' with the issue of so-called 'empty' (in the sense that fictional characters do not exist) names such as those found in novels and poetry (Smith, 2016). We, as readers, tend somehow to 'believe' in the characters and places found in creative writing, but how can something non-existent be tracked through time in the same way as 'real' entities? Berger (2002) provides the most satisfying answer with his notion of mock-focusing and background conditions, which is not unrelated to the idea that real places are being replaced in some way by meaningless or empty places, with similarly empty references (names). Fictional names are, according to this way of thinking, related to or overlapping with (i.e. analogous to) geographical names replaced by quantitative coordinate information, an equally 'empty' construct from the perspective of meaning and (Fregean or Kripkean) reference.

The following, and final, example deals with the idea of a 'final solution' in names and considers whether GIS, GPS, and neogeographical systems (and the like) represent something for places akin to what the holocaust represented for people. These seemingly utterly banal technologies hold the potential to reduce the world to number, quantity, and calculation in a final stripping away of meaning from the world and its places. Space, so conceived, is truly the 'final frontier' to this way of thinking that, taken unreflexively, holds the potential for real and potentially catastrophic remapping of the world (Pickles, 2004). The analogous case, for persons, comes from Levi (1987). If we posit an analogy to Levi's *If This is a Man,* in which the author was stripped of his name and assigned a number, can we similarly posit a 'holocaust of geographical names'? In other words, is the world being reduced to coordinates and neogeographical naming practices where place is stripped of meaning and associated potential for identity formation? In a world without identity/ meaning, can place itself be demonstrated to have, like people, potential agency for the reassertion of difference in the face of such a (hypothetical) holocaust? Is GIS a concentration camp for places (i.e. stripped of names, and reduced to number/quantity)? What are we to make of recent, neogeographical, 'turns' in the geospatial disciplines to revolutionize space and location through the use of words instead of numbers (What3Words, 2016), especially in areas of multiple and overlapping linguistic maps?

It is, perhaps, extreme to characterize such efforts as place genocide. Cleansing and colonization could act as more nuanced frameworks for examining the idea put forward in the last paragraph, using the analogy of Levi and his story of displacement and internment. Similar stories could be told of places (Benvenisti, 2000). The suggestion here is for geographic information scientists (GIScientists) to take up the call and, in essence, to take it upon themselves to make maps of texts, endangered languages, and counter-maps that both contribute to and resist such cleansing and colonizing impulses, tendencies, or trends. This will, at times, prove counter-intuitive or even reactionary. Criticizing (or even critiquing) neogeographical disaster and location mapping will strike many as backward-looking. However, it behoves critical GIScientists to publish ideas critical of well-intentioned and intentionally destructive impulses alike and to remember that 'original' colonial moments of past time were often no less well intentioned (and destructive).

Mapping the future(s) of geographical names

If this book has placed itself in a line of philosophical (Gale and Olsson, 1979) rather than ecological (MacFarlane, 2015; Abram, 1996) thinking, it is no less geographical for that. While toponymy has recently taken a political turn (Rose-Redwood, 2010 and 2011; Berg and Vuolteenaho, 2009), that turn has tended to exclude more critical analytical and/or critical realist thinking in the field of geography itself (with the possible exception of Monmonier, 2006), with toponymic studies relegated to departments of language, including English and History (Nicolaisen, 2001 and 2015; Gelling and Cole, 2014; Higham and Ryan, 2011).

While MacFarlane (2015) and Benvenisti (2000), for example, often focus on landscapes of toponymic loss due to human-induced change, the present work has focused at least as much on toponymic gain. Counter-mapping (Thornton, 2012) and neogeography (Smith, 2008; Pomerantz, 2015) are two areas of development with significance and future interest for the study of names, in any guise, but here I also include anthropology (Turner, 2014) and computing. Human geographical inquiry with GIS is therefore an area to be flagged as requiring more work with the use of geographical names with borrowings from the anthropological and computing science literatures. Mark et al. (2011) remains a touchstone collection in this regard. With the TRC's final report and recommendations release at the time of this writing, the endeavor is all the more pressing (Truth and Reconciliation Commission, 2016).

The bridging of, and blurring of the lines between, human and physical landscapes is, for many overcoming trauma, displacement, or cultural loss, a path of healing. It is hoped that the present work will be added to toolsets for beginning to travel down that (and many other potential) path(s).

References

Abram, David. 1996. *The Spell of the Sensuous: Perception and Language in a More-Than-Human World*. New York: Vintage.

Basso, Keith. 1996. *Wisdom Sits In Places: Landscape and Language Among the Western Apache*. Albuquerque, New Mexico: University of New Mexico Press.

Benvenisti, Meron. 2000. *Sacred Landscape: The Buried History of the Holy Land Since 1948*. Berkeley and Los Angeles: University of California Press.

Berg, Lawrence D. and Vuolteenaho, Jani (eds). 2009. *Critical Toponymies: The Contested Politics of Place Naming*. Farnham and Burlington: Ashgate.

Berger, Alan. 2002. *Terms and Truth: Reference Direct and Anaphoric*. Cambridge, MA: MIT Press.

Doody, Margaret. 2015. *Jane Austen's Names: Riddles, Persons, Places*. Chicago andLondon: University of Chicago Press.

Edney, Matthew. 1993. Cartography Without "Progress": Reinterpreting the Nature and Historical Development of Mapmaking. *Cartographica*. 30(1/2). 54–68.

Evans, Gareth. 1982. *The Varieties of Reference*. Oxford: Oxford University Press.

Gale, Stephen and Olsson, Gunnar (eds). 1979. *Philosophy in Geography*. Dordrecht: D. Reidel Publishing Company.

Gelling, Margaret. 1978. *Signposts to the Past*. Oving: Phillimore & Co.

Gelling, Margaret and Cole, Ann. 2014. *The Landscape of Place-Names*. 3rd ed. Stamford: Paul Watkins Publishing.

Hall, Sarah. 2015. *Wolf Border*. London: Faber & Faber.

Hanna, Patricia and Harrison, Bernard. 2004. *Word and World: Practices and the Foundations of Language*. Cambridge: Cambridge University Press.

Hardy, Thomas. 2009. *Selected Poetry*. Oxford: Oxford University Press (Oxford World's Classics).

Higham, Nicholas J. and Ryan, Martin J. (eds). 2011. *Place-Names, Language, and the Anglo-Saxon Landscape*. Woodbridge: The Boydell Press.

Hornsby, Stephen J. 2011. *Surveyors of Empire: Samuel Holland, J.W.F. Des Barres, and the Making of the Atlantic Neptune*. Montreal and Kingston: McGill-Queen's University Press.

Jubien, Michael. 2009. *Possibility*. Oxford: Oxford University Press.

Kripke, Saul. 2011. A Puzzle About Belief, in *Philosophical Troubles: Collected Papers*. Vol. 1. Oxford: Oxford University Press. 125–161.

Kripke, Saul. 2013. *Reference and Existence*. Oxford: Oxford University Press.

Levi, Primo. 1987. *If This Is A Man*. London: Everyman.

MacFarlane, Robert. 2015. *Landmarks*. London and New York: Hamish Hamilton.

Marechal, Leopoldo. 2014. *Adam Buenosayres*. Montreal and Kingston: McGill-Queen's University Press.

Mark, David M., Turk, Andrew G., Burenhult, Niclas and Stea, David (eds). 2011. *Landscape in Language: Transcdisciplinary Perspectives*. Amsterdam: John Benjamins.

Monmonier, Mark. 2006. *From Squaw Tit to Whorehouse Meadow: How Maps Name, Claim, and Inflame*. Chicago: University of Chicago Press.

Mufson, Steven. 2015. Mount McKinley name change: America's tallest mountain to be renamed Mount Denali by President Obama. *The Independent*. August 31.

Nicolaisen, W.F.H. 2001. *Scottish Place-Names*. Edinburgh: John Donald.

Nicolaisen, W.F.H. 2015. Geographic Names, in Monmonier, Mark (ed.). *The History of Cartography, Volume 6, Cartography in the Twentieth Century*. Chicago: University of Chicago Press.

Pomerantz, Jeffrey. 2015. *Metadata*. Cambridge, MA: The MIT Press.

Pickles, John. 2004. *A History of Spaces: Cartographic Reason, Mapping, and the Geocoded World*. London and New York: Routledge.

Rose-Redwood, Reuben. 2010. Geographies of Toponymic Inscription: New Directions in Critical Place-Name Studies. *Progress in Human Geography*. 34(4). 453–470.

Rose-Redwood, Reuben. 2011. Rethinking the Agenda of Political Toponymy. *ACME: An International E-Journal for Critical Geographies*. 10(1). 34–41.

Smith, Gene. 2008. *Tagging: People-Powered Metadata for the Social Web*. Berkeley: New Riders.

Smith, Grant. 2016. Theoretical Foundations of Literary Onomastics, in Hough, Carole (ed.). *The Oxford Handbook of Names and Naming*. Oxford: Oxford University Press.

Thornton, Thomas F. 2008. *Being and Place Among the Tlingit*. Seattle: University of Washington Press.

Thornton, Thomas F. (ed). 2012. *Haa Leelk'w Has Aani Saax'u/Our Grandparents Names on the Land*. Seattle, London and Juneau: University of Washington Press/ Sealaska Institute.

Truth and Reconciliation Commission. 2016. *Canada's Residential Schools: The History, Part 1, Origins to 1939: The Final Report of the Truth and Reconciliation Commission of Canada*. Vol. 1, Part 1. Montreal and Kingston: McGill-Queen's University Press.

Turner, Nancy J. 2014. *Ancient Pathways, Ancestral Knowledge*. Montreal and Kingston: McGill-Queen's University Press.

What3Words. 2016. The Simplest Way to Communicate Location. what3words.com (accessed January 15, 2016).

Wordsworth, William. 2008. *The Major Works*. Oxford: Oxford University Press (Oxford World's Classics).

Yeazell, Ruth Bernard. 2015. *Picture Titles: How and Why Western Paintings Acquired Their Names*. Princeton: Princeton University Press.

Index